Ponderings of a PPC Professional
(Revised & Expanded)

KIRK WILLIAMS

Copyright © 2024 Kirk Williams

All rights reserved.

ISBN: 979-8-9860306-4-7

DEDICATION

This book is dedicated to the Fab 4 of ZATO.

Eric, Sarah, Diana, and Chris, you never stop pushing me to learn more about PPC and to be a better paid search practitioner, speaker, author, employer, and person. Working with you is truly a delight each and every day, and I consider myself incredibly blessed to call you team members as well as friends. Please never stop challenging and questioning me.

Onward & upward, if not always in revenue, at least in PPC learnings

Ponderings of a PPC Professional: Revised & Expanded

CONTENTS

	Acknowledgments	i
	Preface	1
	Introduction	4
1	The Keyword: Personal, Timely, Intent	7
2	Keywords Aren't Just Topics	12
3	The Concerning Future of the Resilient Keyword	17
4	Controlling the Keyword for Small Budgets Is Crucial	28
5	To Bid or Not to Bid on Brand	32
6	Invest in More Channels Than PPC or Die	38
7	When to Blame Poor Positioning, not PPC	44
8	The Marketing Funnel Is Not Dead	48
9	When Optimizing Ads Strangles the Funnel	54
10	PPC Isn't Primarily for Scaling	61

11	Balancing Growth with Profit Over Time: The Ascending Seesaw of Scaling	63
12	Eleven Reasons Your Google Ads CVR May Have Dropped	70
13	A Tale of Attribution Woe and Cow Crap	76
14	Avoid These Two Ditches of Digital Attribution	79
15	The Digital Attribution Bubble	84
16	When Data & Automation Disagree	88
17	Testing PPC With Low Budgets	97
18	Let's De-Frenzy PPC	103
19	The remarkable and alarming power of the Google Ads Platform	114
	Appendix A: An Open Letter to Google	122
	Appendix B: Pros & Cons of PPC Agency Pricing Models	130

ACKNOWLEDGMENTS

I couldn't do this without my ZATO team, but I also couldn't do this without my wife, best friend, and life partner. Elise, I am a better person for knowing you, and that makes me better at my work as I pursue deeper understanding of PPC.

I also couldn't do this without my children, the 6Bs, who constantly challenge me to pursue deeper learnings in my work, even as I challenge them to the same in their education.

Thank you again, to my editor (and the best Mother-In-Law in the world!) Cindy Allen. Your tireless willingness to call out my dangling prepositions is something I "aspire to", and "than" I work on fixing "more then" I was capable of without your help.

I would also be remiss in not thanking the entire ZATO, DTC, and digital marketing world. Whether you are taking shots at me in your well-thought-out emails and LinkedIn posts (I see you Ben Kruger), nodding your head in support (I see you Sarah Stemen), or cheering me on in one comment and pushing back with an alternate position in another (I see you Navah Hopkins), I wouldn't be who I am, and this book wouldn't be whatever it is without you. Thank you for being the best business community out there!

PREFACE

Balance isn't exactly something that we always have in excess, at least in the world of digital advertising. Since we're all just humans (at least I hope so!) it's easy for us to make knee-jerk reactions, foretell doom, and prioritize short-term gains over long-term prosperity within our industry.

This is precisely why Kirk's perspective is refreshing for the paid search world: he tells a much-needed story of balance.

In early 2023, Anya and I met up with Kirk in San Jose. We'd set up some casual talks at Google's new headquarters on the bay. As we sat down with Google Ads product managers and talked about everything from automation, to attribution, to reporting, to machine learning, to the nebulous future of the keyword, I discovered that Kirk's approach to PPC was shockingly similar to our own. He's got a very fair & balanced approach: not so much of knee-jerk "Google bad!" or "automation no good!" reaction, but rather a more measured "how do we make this work for our clients?"

Reality is nuanced. And that's why I'm excited for you to read this book and discover some of the same nuances that will help you run your campaigns in the most effective manner possible. Despite the constant flux of the digital advertising ecosystem, the reality is that with every change we don't like, there's also a constant stream of exciting new developments that we would be remiss to ignore.

Kirk's balanced approach isn't just relegated to PPC, but

to business in general. If you've read Kirk's other book, Stop the Scale, he poses thought-provoking questions that every marketer or agency owner should ask themselves at some point: how much is too much? Is growth the ultimate goal? Who are we working for? Questions like these not only shape how Anya and I manage our PPC agency but also mirror the way we manage Google Ads campaigns.

This balance becomes really important when we're down in the weeds of PPC. As it turns out, focusing on long-term success, building sustainable business practices, and getting back to the roots of marketing works well not only in creating an agency you'd like to work at, but also in creating marketing campaigns you'd like to run.

Gil Gildner
Co-founder, Discosloth
Author, Building A Successful Micro-Agency and Becoming A Digital Marketer

INTRODUCTION

It is 2024, four years after the beginning of the Pandemic, and four years after I wrote my first book, PPC Ponderings. As no surprise to anyone, the changes in both the world, ecommerce, digital marketing, and PPC have been significant... perhaps even accelerated, based on the chaotic coinciding of those two categories. My digital agency focuses primarily on the Ecommerce world, and even four years later, the influence of the pandemic continues to send specter-like shivers down the spines of tired brand operators.

From wild shipping container price swings, to digital security changes, to consumer purchase behavior changes due to rising inflation, throwing in a dash of political unrest and the sudden explosion of Chinese apps galore (TikTok or Temu, anyone?), ... the Ecommerce Digital Marketing world is tired.

It is within this world that I considered the reality of the likely out-dated nature of 4 year old ponderings in my previous book, and I determined it was time to rewrite, add, and update my thoughts so we could continue to ponder together.

My goal in this book is to maintain the core of my original intention in writing PPC Ponderings, but to do so with 2024 realities. As much has changed in the broader world, that much (and perhaps more) has changed in Paid Search. When I wrote the original PPC Ponderings book, Performance Max campaigns were still just a twinkle in Google's collective eyes, exact was still kinda exact, iOS 14.5 had not yet occurred, and 3rd-party cookies were all the rage. Yeah, now that your eyes have widened, you understand why it was time to revisit a book on PPC. Some of my ponderings and chapters will remain unedited. Some will remain and get updated, some will be removed entirely, and new chapters will be added with my latest ponderings since 2020.

Let me be clear, however, this book will continue to be more philosophical than tactical. If you are looking for some book to give you the four secret hacks of PMax strategery, this is certainly not it. I mean, by the time you pick this up, Google will likely have changed something so significant that any tactics that worked in 2023 are already outdated. The point of this book, as was my original PPC Ponderings, is to push us to think more about the philosophies that drive Paid Search, and thus allow us to make our own individually solid tactical decisions based off of a solid philosophical foundation. In other words, for all that is good and holy, please don't write an Amazon review bashing the book because it didn't tell you which smart bidding model is best.

Whatever the result of this finalized book, I'd like to add what I included in my original Introduction at its close since it is with this intention, I again put metaphorical pen to paper:

I love the word "pondering." It means to think carefully, to mentally chew through something over time. The writing I do in PPC often results in something I have pondered on, sometimes for years. I wanted to write a book based on those various ponderings, and in doing so, encourage you to embark on your own pondering experience on these various aspects of PPC. While there are excellent books about the history of Paid Search and the inner workings of how remarketing audiences work and should be added into an account, I wanted to write a book about the random, important, philosophical things I think are necessary to know about PPC. These are the random things that come to my brain, the important things that you might not hear elsewhere, but that will make you a better PPCer because they will get you to think about how PPC works.

My goal is to challenge some commonly held beliefs and to support others, but above all… to get you to think. If, by the time you finish this book, you have not disagreed with at least one thing in the book, then I have failed you. I am not a perfect marketer, I will have gotten some things incorrect (which frankly, makes putting myself out there quite terrifying), and my goal isn't to get you to agree with everything I say. My goal is to get you to think. If I accomplish that while you read this book, then I can live with the knowledge that I don't have it all correct myself, because I have helped us think through something just a little more deeply than we would have otherwise. Your attention is a privilege I do not take lightly. It's a brave new world out there in the Land of PPC, so let's start exploring and see where this adventure takes us both.

Kirk Williams
January 2024
Billings, MT

1 THE KEYWORD: PERSONAL, TIMELY, INTENT

It is 2024.

It is four years after I first wrote these words about the keyword being the true power of Paid Search, and as I pondered which chapters should remain in the updated version of this book, I thought it necessary to leave this one in.

We'll be discussing the keyword a lot in this book, especially in terms of changes we should be considering. But at its core, my belief is that the idea of the keyword itself, from a marketing perspective, is largely unchanged. Now, Google has changed our ability to target or bid on specific keywords, but this doesn't change the philosophy of the keyword, and I think it only fair we begin once again by taking a deep look at the keyword.

Keyword targeting does something audience or demographic targeting cannot do… it reveals (1) individual, (2) personal, and (3) temporal intent. Those aren't just three buzzwords I pulled out of the air. They are intentional and worth exploring.

Individual

A keyword is such a powerful targeting method because it is written (or spoken!) by a single person (let's be honest, it's rare to have more than one person huddled around the computer shouting at it outside of a livestream of your favorite sportsball event). Keywords come from the mind of one individual (grouped together, sure, but they begin as individual entities!), and because of that they have frightening, personal potential.

Whereas, keywords reveal personal intent, audience targeting is based on grouped assumptions. You're taking a group of people who "probably" think the same way in a certain area based on limited knowledge you have of past behavior (made even more difficult to track accurately in a shared device, and privacy-obsessed new era of digital). But does that mean they cannot have unique tastes, such as one person preferring to buy sneakers with another preferring to buy heels? Imagine twin sisters who appear to be the exact same by every demographic average, but they have very different personal interests.

Keyword targeting is demographic-blind. To paraphrase the Backstreet Boys, It doesn't care who you are, where you're from, what you did, but what you're individually interested in (and what you communicate directly to Google).

Personal

The next aspect of keyword awesomeness is that they reveal personal intent.

Whereas the "individual" aspect of keyword targeting narrows our targeting from a group of people to a single person, the "personal" aspect of keyword targeting goes into the very mind of that individual. Don't you wish there was a way to market to people in which you could truly discern the intentions of their hearts? Wouldn't that be a powerful method of targeting? Imagine if you could determine which of the 200K people who drive by your billboard are interested in your ad, and only pay for them? Wouldn't that be an unbelievably powerful form of advertising?

Well, yes it would be, and that is keyword targeting!

Think about it: a keyword is a form of communication. It is a person typing or telling you what is on their mind. For a split second, in their search, you and they are as connected through communication as Alexander Graham Bell and Thomas Watson on the first phone call. That person is revealing to you what's on her mind, and that's a power that cannot be underestimated.

When a person asks Google: "how does someone earn a black belt," they are telling you (the local judo guru) that they genuinely want to learn more about your services. You can then display an ad directly to them that matches their intent.

Ready for that Black Belt?
Anyone Can Take Our Classes. Even You.
JustJudo.com

Paid search keywords officiate the marriage of advertising to personal intent in a way that previous marketers could only dream. We aren't finding random people we think might be interested based upon where they live. We are responding to a person telling us they are interested.

Temporal

A third note of keyword targeting which cannot be underestimated is the temporal aspect.

Anyone worth their salt in marketing can tell you "timing is everything". With keyword targeting, the timing is inseparable from the intent.

When is this person interested in learning about your Judo classes? At the time they are searching, which is NOW! With the keyword (i.e., Search Advertising), you are not blasting your ads into your users' lives, interrupting them as they go about their business or family time, and hoping to jumpstart their interest by distracting them from their activities.

You are responding to their query at the very time they are interested in learning more.

Timing. Is. Everything.

Demand Capture

There is a final aspect of the keyword that is more caveat than clarification. While the keyword is individual, personal, and timely, it is important to understand the place the keyword thus sits within the marketing funnel. In order for a person to formulate a thought from which to search, that person must be aware of the concept for which they are searching.

Because of that, paid search (keyword targeting) tends to fall into the demand capture realm of marketing. This is separate from the concept of demand generation, in which a person going about their lives suddenly learns (from the insertion of an ad, typically in a way that distracts them from their current state) of a new product. Paid Search, requiring a person to put their thoughts into words, requires pre-existing knowledge of the thing for which they are searching. This gets messy, of course, since people can certainly search in general categories for which an advertiser can "hijack" the intention of the search, though I would argue there is still a "demand capture" element occurring.

Let's say a person types into Google "best headache remedy" and you sell them a coffee subscription. They weren't technically searching for "coffee", you just happened to advertise on an adjacent search term. But even then, I would argue that they were actively searching for a solution to a problem, which carries a mid funnel quality to it.

I bring this up because it's important to realize the keyword has its own restrictions in the midst of your overall marketing strategy. Is the keyword brilliant? Is it an amazing form of advertising? Absolutely, but a business needs more than simply search marketing in order to scale, since you can only capture so much demand, especially in a profitable manner.

This brings us up to another aspect I want to discuss before closing out this chapter, which is the cost of the keyword.

Cost Considerations

While I have trumpeted here the glories of the keyword, I will now step into the realm of reality. In 2024, keyword clicks are more expensive than they used to be. There are many reasons for this, not solely because an auction system will naturally have some increase to it over time as it matures. However, the savvy marketer will not put all their eggs in the basket of Paid Search due solely to the fact that the keyword is a powerful form of marketing, as I have argued above.

It is possible that someone may simply not have the budget to capture the demand, albeit individual, personal, and timely, for which their industry is priced. For example, Legal and SAAS terms are known for being very pricey. It could be that a startup SAAS company should consider not investing immediately in an exhaustive search strategy in the beginning, not because the keyword itself is lacking, but because their specific vertical may mean it is wiser for them to invest the high expense into influence, or social marketing.

This section should not be considered an exhaustive strategic resource for how to begin marketing a business, but just a brief caveat to note that a keyword's incredibly targeting ability does not always make it a wise choice for a business, though there are certainly ways a savvy PPCer will know how to make a budget last for smaller budget accounts.

It ain't 2018 out there when it comes to keyword CPCs, that's for sure.

In Summary

So, let's summarize: a "search" is done when an individual reveals his/her personal intent with communication (keywords/queries) at a specific time.

Because of that, I maintain that keyword targeting is still an incredibly powerful method of ad targeting.

Paid search is an evolving industry, but it is still "search," which requires communication, which requires words (until that time when the emoji takes over the English language). Thus, it is my belief that as long as we have search, we should still have keywords. Keyword targeting has been and will be the best way to target (as long as costs remain low enough to be realistic for budgets and the search engines don't kill keyword bidding for an automated solution).

Don't give up; the keyword is not dead.

2 KEYWORDS AREN'T JUST TOPICS

Over the past few years, I've pondered the idea of Paid Keywords and their role in marketing, as well as specifically how their targeting has changed within the Google Ads ecosystem. That should be clear by the keyword focused content in my first version of this book, and in this one as well!

Based on this, I'd like to push back on a statement I hear quite a bit, and the argument goes something like this:

"Well, keywords don't really matter because they are just topics anyway."

This argument typically is given after some sort of discussion on close variants, the usage of broad match keywords, the expansion of audiences within Google Ads, or even the utilization of Performance Max campaigns in an account.

There is much I agree with in the above statement, and it would be helpful to begin with that.

I believe what is meant when a variation of this phrase is uttered is that we can't be so rigidly obsessive over specific phrases that we forget that people overall mean different things by what they say. What's more, people can even say different words to mean the same thing. So in that sense, the order of words, and even at times the specific usage of certain words, does not necessarily equate perfectly with intent.

As an example, people asking Google for the following three keywords are all, arguably, wanting to see the same ad.

"best treadmill desk"
"most durable treadmill desk"
"a treadmill desk that won't break on me"

In this case, let's assume their intent is to view an ad for a higher end treadmill made with more commercial grade materials that works for them in a long-term capacity. In this case, the searchers have different starting places but really are speaking about the same topic, and Google is, arguably, correct in seeing these three (in that specific use case) as sharing a topical interest.

I can agree that there are variations of the above example, in which the specific words in a search term may not always matter quite as much.

All that being said, here is where things get muddy and why I think the above statement tends to be an unhelpful oversimplification of the Paid Search world of targeting.

Keyword targeting isn't JUST about topics, it's also about intent.

Someone may be interested in the same topic, but have very different "intent". This is where things like close variants can really fall apart, and why I've argued before for a return to making "exact match be exact". By this, I simply mean allowing for expansion by Google as budget allows in the phrase and broad matches, but allowing for a purposefully restrictive match type like Exact to remain for more advanced marketers. Yes, even for misspellings and plurals, for which Brad Geddes has previously made a case.[1]

In other words, I think Google should treat Exact Match like Negative Keywords are treated within its advertising matching

---

[1] Brad Geddes, "AdWords is Forcing Variation Match Upon You-This is Why it's a Terrible Idea," blog post, 2014, *BGTheory Blog*, accessed 31 January, 2024, https://bgtheory.com/blog/adwords-is-forcing-variation-match-upon-you-this-is-why-its-a-terrible-idea/.

system. Negative keywords do not utilize close variants, an Exact match negative keyword must exactly match the search term before it excludes the ad from that search, and a Phrase match negative keyword must specifically match the phrase searched for in order to exclude the ad from that search.

Actually, pause to ponder that point a bit. If Google truly believed that close variant behavior was the best way to manage targeting (as the best possible experience for the advertiser), wouldn't they also apply close variants to your negative keywords as well? If their highest objective (as they note for a keyword close variants defense) is to allow for helpful targeting shortcuts based on semantic connection between various words... wouldn't they also apply that same logic to negative keywords that would "help" an advertiser be more restrictive to their specific topic/intent? Clearly there is more at play here for Google than simply finding the best semantical matching, and we would be wise to keep that in mind with close variants as well.

That being said, words certainly have a topical nature to them (as we discussed earlier), but we can identify actual intent-based trends based on historical performance, which is arguably more significant in bidding and organization than is a topical nature alone.

In other words, the exact match keywords: [contractor license] and [licensed contractor] appear to be the same word with the order switched around, but they have very different meanings (see the old but still insightful article by Brad Geddes: "Is Phrase Match Dead?"[2]) and Google doesn't always get this correct (to which any of us with Search Terms Reports can attest). There are even times when misspellings and plurals show dramatically different purchase intent by users.

Why?
I don't know why.
I don't know if we can always know why, and that's exactly the point. People don't always make sense, but we can still set unique

---

[2] Brad Geddes, "Is Phrase Match Dead?" blog post, 2016, *BGTheory Blog*, accessed 31 January, 2024, https://bgtheory.com/blog/is-phrase-match-dead/.

bidding targets and create unique ads for semantical phrases with matched purchase intent based on historical performance. In other words, if an exact match term converts better than another, even if they are very close in supposed intent and they clearly are within the same "topical realm", it still behooves the wise advertiser to give those terms dedicated bids/budgets/ad text/landing pages to eke any remaining difference in performance they can find.

The way I see it is, in that instance, the specific arrangement of words IS the audience (I'm going to define that here with my own made up definition: "a group of people to market to with shared characteristics"). We have to stop pitting audiences against keywords as if they're two separate entities. Sometimes the keyword needs to be seen as the "audience" based on historical performance, and then we can go about optimizing. That's old school Paid Search, but that's because it matches the beauty of marketing keywords. Keywords aren't just semantical phrases. We can't assume we know what a keyword will be simply because it appears to be high intent, and that's why keywords still matter as unique targeting entities.

They are far more complex than simple topics. They are living, breathing marketing targets that have topic and demographic and intent based activity within them, and sometimes the very best way to manage them is to look at performance, and then pull them out into separate entities and optimize the snot out of those individual terms. They can even change over time, as the search behavior and language of people changes.

This is especially true with limited budget Google Ads accounts. I still manage my wife's local photography business ad account ($250-500 in advertising spend per month) successfully by focusing almost solely on Exact (and some phrase) match keywords (while keeping up on negatives!) because exact match keywords are (still) fantastic for limited budgets!

"But you have to trust the machine!!"
Someone may shout in a last ditch effort to call me a luddite for having the gall to use Exact match terms in 2024. To that, I'll reply: We've been trying to trust the machine, but our search terms reports continue to get in the way of this trust.

Just yesterday I was in an account preparing a proposal for a prospect and pointed out that Google was close-variant matching a NON-brand (very!) generic term to their decidedly BRAND exact match keyword (and it was not a generic brand, like "Best Home Tools dot Com" or some other keyword stuffed name).

Why would Google do that? You might surmise: perhaps the keywords performed at a similar level so Google is finding high value customers typing in similar-intent terms?

Nope, not in this account. The non-brand keywords were dramatically underperforming historically in the account while the Brand keywords were performing at a high Return on Ad Spend level. There was enough data in the account to not match this search term. Machine Learning failed. Watch social media long enough and you'll see case after case of this.

Here's the problem. It's not that Google always fails... it's that it fails enough, in enough accounts, to really mess up the targets and budgets of a lot of accounts. In many cases like this, the business owner doesn't even notice because they don't really understand what's going on, so they could be overbidding for those "generic" terms that are close variants for the Exact Brand keywords since those terms tend to be higher bid than generic non-brand keywords.

That's why it's an actual issue. Because of close variants in those cases, the advertiser ends up spending more than they would have otherwise while advertising those same generic terms elsewhere in the account at lower bids... and that's why it's difficult for us PPCers to trust Machine Learning. We've seen too much stuff like this.

So, rambling soapbox down, all that to say, I think it's helpful to view keywords as having a topical element. But with that in mind, there is also a purchase intent that needs to be taken into account, even when topical behavior is shared across multiple search terms... and close variants can struggle with this.

Because of this, I think there will always be a place for exact match keywords if Google will allow for them. Let's make Exact match Exact again for our core terms and use Broad, Demand Gen, and DSA to find new opportunities elsewhere in the account.

3 THE CONCERNING FUTURE OF THE RESILIENT KEYWORD

It is with tears (metaphorical) that I write this chapter. I weep not solely for my loss, but yours as well. I am referring, of course, to what I see as the potential demise of a legend, an icon, the darling of all marketing land:

The keyword.

You may be shocked after reading the last few chapters in which I trumpeted the beauty, power and strength of the paid search keyword. Why the sudden change of heart?

Well, I think there is a difference between what the keyword is capable of and what Google will change over the next few years. If you ponder the degradation of the keyword with close variants, and a push towards completely keywordless campaigns such as Performance Max and Demand Gen Campaigns, I would not be surprised if PPC advertisers no longer have the ability to bid on individual keywords in the next few years.

That being said, I also find myself in the unenviable position of changing parts of my viewpoint. "Changing" is a strong word; perhaps adjusting or tweaking is a better term here?

The remarkable power of the keyword

Make no mistake that I still believe the keyword to be the most remarkable marketing tool known to humankind. Let's review what we've already learned in this book:

A keyword is an advertiser's means of selecting which *search terms* they want to advertise on. The search term tells you exactly what individuals want, and when they want it. This gives advertisers the opportunity to position their product or service in immediate response to that communication, and to even determine what value (the bid) that exchange carries. Absolutely remarkable!

Aside from someone physically walking up to a marketer and asking them a question (we typically call that "sales"), where else can this sort of immediate connection be seen in the world of marketing? This is why we PPCers have always battled changes to the keyword.

"Keep exact match, exact!" is an oft-repeated battle cry heard over the past few years on Twitter and in PPC conferences (for the youngsters: conferences were an impressive conglomeration of people who could actually communicate in person, back in the day).

If Google Search is a channel built on language, words, semantics, then it makes sense that the power to target ads to language, words, semantics would be an aligned method of advertising within said channel. Words matter.

But, I am also willing to consider other aspects of communication and Google as I ponder this question: have times changed?

In the remainder of this chapter, I hope to reexamine the PPCer's traditional understanding of the keyword by presenting new information I believe should not be ignored.

Yet, I also want to walk a fine line: For anyone who has followed my writing over the past few years, you can relate to the fact that I am no Google shill, buying into their every edict with breathless anticipation. While I am happy to learn Performance Max (and consider advanced ways to manage this campaign type), I have also been a vocal critic (on multiple social media platforms) about the alarming data loss in PMax and its harm to businesses and advertisers.

On the other hand, I want to be cautious not to criticize everything Google does immediately without all the facts. That, of course, can get difficult when the company withholds information from advertisers making it seem suspicious... admittedly, sometimes I believe that has more to do with poor internal communication than we conspiratorial mutterers tend to always admit.

Anyway, let's get to it. What has changed in my understanding of the keyword?

A potential new weakness in the keyword

As much as I love the keyword and still believe it carries immense value, I have also begun to question whether it is still telling us as much as we think it does.

Understanding this is important for our industry to move forward because, if the keyword is not telling us as much as we think it is, then we have an obligation to evolve with Google's system for the good of the accounts we manage. After all, we're here to grow our PPC accounts the best we can; that is our ultimate goal: to serve our clients or place of employment well.

What I would currently argue is that the combination of changes in user behavior, changes in Google Ads automation and targeting, and an increased number of privacy changes have made the keyword less of an intent-revealing ace-in-the-hole than it used to be, though certainly still a powerhouse in its own right as I have argued in previous chapters.

Because of that, we must also accept user-defined targeting as a legitimate additional targeting method in order to fill in the gaps of our contextual knowledge of searcher intent, whereas in the past, we could rely almost solely on the keyword.

Lest your eyes glaze over, let me get practical with an example before moving into the details: A user is interested in purchasing a new TV. They go to Google to type something into the computer. How should we bid on them?

Advertisers' changing relationship with the keyword for targeting

Circa 2015 PPC:
In the old days, we would build lists (I'm talking, LISTS) of detailed long-tail keywords (exact match, of course!) based especially on modifiers (where we really could flesh out intent).

This might look like thousands of ad groups according to the following:

[best flat screen tv under 1000]
[65 inch plasma tv above fireplace]
[samsung large lcd tv]
[samsung tv vs sony tv]

While we could use broader keywords for upper-funnel targeting, the truly savvy advertisers got focused and ultra-specific with what terms they would target. Long live the long-tail, purchase-intent revealing keyword!

PPC in 2021 and beyond:
Privacy regulations and changes are beginning to wreak havoc on the specificity of tracking. In addition, user behavior has been trained by Google personalized results to never spend more time than they have to when typing (or speaking) into a phone (more on this to come below).

A few years ago, Rand Fishkin gathered data and presented some remarkable findings on search behavior trending towards shorter queries.[3] He discovered that 46% of searches are one- or two-word search queries. Nearly one out of every two searches doesn't even get to three words! Imagine what that is now (no,

---

[3] Rand Fishkin. (2017) "How Marketers Can Keep Up With Google in 2017 and Beyond," Slideshare, accessed 31 January, 2024, https://www.slideshare.net/randfish/keepng-up-wth-seo-n-2017-beyond#11.

really, I'd love some data on this... I looked and asked, and I couldn't find any).

Tim Soulo of AHREFs reminds us that long-tail keywords are not the same as multiple-word keywords (a good reminder) in his helpful and accessible article on the long-tail keyword.[4] It is helpful to note his keyword length vs. monthly search volume chart within the article, which also demonstrates the significant number of fewer-word keywords in the head terms (high search volume) camp.

While users still search for a significant amount of new, long-tail keywords (regardless of word length within the term), my hypothesis is that user behavior has changed, along with platform changes, to result in a different PPC keyword landscape than has ever been seen before.

The wise PPCer will identify the ways in which adaptation is important, even while frustrated about losing search term data. We are stuck now, because the best-converting keywords to bid on in our accounts have likely shifted (in many, but certainly not every, instance) to look like this:

[best tv 2021]
[led tv]
[samsung]

If we had bid on those three terms in 2015, our budgets would have resembled waste flushed down a toilet, and disappearing quickly. Now, those terms may be the core of our accounts.

Here is my assertion: We are seeing a shift in user behavior to intent-obfuscating search terms, which then impacts purchase behavior and our keyword targeting and bidding capabilities. This shift (along with Google Ads changes that may or may not be in reaction to this shift) is going to continue to force us to rethink our keyword targeting strategies.

---

[4] Tim Soulo, "Long-tail Keywords: What They Are and How to Get Search Traffic From Them," blog post, 2021, *Ahrefs Blog*, accessed 31 January, 2024, https://ahrefs.com/blog/long-tail-keywords/.

Why is this happening?

This is a fairly remarkable change, and I think it's worth considering why it has happened. If I might be allowed to ponder publicly (which runs the risk of me being wrong), I'd suggest four key reasons why this is happening, and it's important to note they are all interconnected. These are observations formed to create an assumption. I could be wrong (but hey, you could be, too, right?) so I hope this at least encourages fruitful discussion.

Here is what I believe has provoked a change in how we should view keyword targeting in PPC. I should also note that internal conversation on our team has helped immensely with this as well. I have a brilliant team, and they've helped suggest or hone the specific points below.

#1. A better product. I have to hand it to Google, as the company has done a remarkable job with its search results. Think about it (including your own search behavior): When people go to Google now, they experientially know they don't need to fill in all the gaps because Google personalized search seems to somehow just... know them. People have been increasingly trained by Google (without knowing all the details of their previous behavior being tracked to build that product) to get lazier in their searches because Google's results will still be personalized to them. That's pretty remarkable, and Google has indeed built an impressive search product.

Example:
Let's say Person A is in the market for a mattress and has spent weeks investigating numerous brands online, researched local mattress shops, clicked on specific shopping ads to view individual products they're interested in. Google has all of this data, and the next time they go to Google ready to purchase, they simply type [mattress] and Google knows exactly which Shopping Ads (let's say, in the form of dynamic remarketing within a Performance Max campaign) or local stores they should show to Person A.

Person B, on the other hand, is just beginning their shopping experience online, so when they type [mattress], Google changes the results to a higher-funnel, more informational SERP (Search

Engine Result Page).

Same keyword, dramatically different results, amazing product!

#2. Autocomplete (plus constant improvements). Along with the first point, consider how Google autocomplete plays into a change in user search behavior as well. I would go out on a limb and say I now use autocomplete more than not. Autocomplete really is remarkable in its accuracy (though, humorously not at times) and this goes along with the better search product mentioned above.

I've even gone to Google.com at times not knowing exactly what I will type (and not caring that I don't know), since I know Google autocomplete will figure it out. Again, credit to the impressive product Google has built, but autocomplete certainly impacts how people search in bulk.

One of the things Tim Soulo brought up in his article (that I previously linked to) is the surprising number of many-word terms searched for in head term searches. According to Ahrefs' data, 9% of all searches with 10,001+ monthly searches have four or more words! That is remarkable, but I wonder how much of that is influenced less by synchronization and more by autocomplete.

In fact, while writing this chapter, I did a search on Google and used autocomplete out of habit and ended up clicking on a five- or six-word term. If Google senses trends in specific phrases and raises those in autocomplete, it stands to reason that act will, in itself, increase the number of searches done for that exact phrase.

That changes the nature of how people search and the value of certain keywords for bidding within Google Ads.

#3. Mobile device impact. What would a change over the past years in user search behavior be without the impact of mobile? What would a chapter about digital marketing be without mentioning mobile?

If you remember such declarations as "Mobilegeddon" a few years ago, there was a shift of Google searches from desktop to mobile. This never fully transitioned as proponents of the concept suggested, simply because there are still a healthy number of people who utilize (and will likely for years to come, in my opinion) desktop for things like work.

Yet, mobile devices have an impact on search behavior, in part, due to practical considerations such as limited screen size and inefficient touchscreen keyboards (I must be getting old, because typing on an iPhone keyboard still makes me want to throw my phone into a blender).

One dataset I have never seen, but would like to, is whether the number of words in a search term tends to decrease based on device. I would be fairly shocked if it didn't due to the aspects I noted above. It is also possible that "more-word" terms tend to be more voice-centered while "fewer-word" terms tend to be input via touch screen. This is admittedly a calculated assumption based on known behavior rather than fact based on gathered data.

#4. Google Ads close variants. Finally, we get to the Google Ads product itself. I purposefully left this as a later point because I think we PPCers have tended to underemphasize the influences of the factors mentioned above on changing keyword targeting.

However, we would be remiss in not acknowledging the changes within the Ads platform itself on keyword targeting — one of those being close variants. Is this a form of Google forcing an outcome with its platform decisions? Absolutely. But then that carries with it the practical weight of adapting to existing dynamics... even as we push for better engineering efforts (as I have tried to do in other parts of this book).

Anyway, with close variants, Google has made exact matching of long-tail (as well as more-word) search terms to specific advertiser-selected keywords more difficult as contextual and close variants are now considered. Negative keywords may assist in filtering out some of the unwanted variations into a preferred ad group (as some advertisers have developed into "query filtering" or "query funneling" strategies); however, traffic availability may not always allow for this... it can sometimes actually prevent your overly-filtered keywords from serving, rather than successfully filtering the term into the correct keyword within your structure.

This has certainly had an impact on the changing nature of keyword targeting. Google argues that this is to allow the advertiser to match up to the many searches (mentioned earlier) that are contextually the same as the keyword selected, but that do not match it exactly and would thus be missed in the advertiser account. If it were up to me, I would prefer broad and phrase match to allow for close variants while leaving exact match a level of exact specificity. This would (I believe) unite the role of machine and human oversight in keyword targeting in a healthy manner, especially for budget and ad creative control to core terms.

But here is where I think it's crucial to reassert the purpose of this chapter: I would point to the first three reasons above as to the cause of a shift in user search behavior. We PPCers can be so focused on the data points within our ads system that we miss the human aspect here, which also drives our targeting capabilities.

Yes, the close variant matching has had an impact on our ability to target as specifically as we were able to target in the past. But I would argue that we are also seeing a shift in user behavior to intent-obfuscating search terms, which then impacts purchase behavior and our keyword targeting and bidding capabilities. This shift (along with Google Ads changes that may or may not be in reaction to this shift) is going to continue to force us to rethink our keyword targeting strategies.

Bonus: #5. Privacy. I would have failed if I didn't at least throw a passing nod to a core aspect of the search environment that is in constant flux: privacy and security.

As Google continues to eliminate the third-party cookie and to institute its own new initiatives for tracking (along with first-party cookie products, of course), users are certainly more aware of data, tracking and their own privacy than they have been in the past. I do not have anything to point to quantifiably here, other than to suggest something as socially contextual as user privacy likely plays a role here in changing user search behavior.

Practically, more people than ever have been shifting to privacy-encouraging browsers such as Brave over the past few years, and news articles such as Google being sued for tracking users within their Incognito browser window[5] suggest the average internet user is more aware of and savvier regarding privacy decisions. It isn't too much of a stretch to assume that this could work its way into search behavior as well. However, I don't think it is as significant as the first four, so I only noted it less as a root cause of a clear user

---

[5] Malathi Nayak & Joel Rosenblatt, "Google Must Face Suit Over Snooping On 'Incognito' Browsing," blog post, 2021, *Bloomberg*, accessed 31 January, 2024, https://www.bloomberg.com/news/articles/2021-03-13/google-must-face-suit-over-snooping-on-incognito-browsing.

shift in behavior, and more as something to be aware of and monitor.

So what do we do about this?

I think the key is to evolve with this change in user behavior and the Google Ads platform changes by adapting to a more user-targeted system and shifting strategies and tactics accordingly.

Think about it like this: If 1,000 people are shown your ads for "samsung tv," and, in this paradigm of new user behavior that I have outlined above, 20% of those people are ready to purchase within the next seven days, 20% are in the education and shopping phase, and 60% are too high in the funnel to determine a strong purchase intent, then don't we need additional context in order to capture the 20% that are ready to purchase now?

Here is where the additional user signals (that Google alone has access to) become essential. A frustrated PPCer may exclaim that Google shouldn't be the only entity with access to that data (I am in agreement with this concern as I have published previously), but that is an argument for a different time since I am trying to pragmatically explain how I believe PPCers need to evolve within a changing system based in large part by a change in user search behavior.

Google has stated that "millions of signals" are used to determine bids for an advertiser within a single auction.[6] So how does Google know whether that person typing in a generic two-word query is closer to purchase or not? Because they have a mind-blowing number of data points they have collected on that user based upon things like previous search behavior, sites visited, affinities and interests and a host of other behaviors to which we humans simply do not have access.

Legitimate concerns about closed systems aside, this is one of the strongest arguments I have heard for embracing more automation within Google. I simply don't have access to that level of data in a real-time auction scenario, and with decreased intent

---

[6] Google, "About Smart Bidding and Smart Creative solutions with Google Ads" Google, accessed 31 January 2024, https://support.google.com/google-ads/answer/9297584?hl=en.

revelations based on the keyword itself, I also don't have a lot of other options. In the words of Patrick Gilbert (and title of his must-read book[7]), the PPCer of the modern age must "Join or Die." That's not a threat (at least from me or Patrick), it's a warning.

A plea, even.

What does this mean for the keyword itself?

It was my friend, Aaron Levy, who once referred to the keyword's future as suspect simply because the idea of the keyword has already changed. "Think of the keyword more as topics or categories," Levy noted (I paraphrased) in a PPC Ponderings podcast episode on the keyword.[8] I think Aaron has nailed it, and this supports what I have been outlining above.

I believe the wise PPCer will begin to prepare for a world in which the anonymized and aggregated data points on individual users (which only Google has access to and can utilize) are the primary signals used, and the keyword is simply one part of this system (albeit, one in which significant weight should be placed based on the nature of its intent-wielding power, as outlined in the first part of this chapter).

Lest my previous chapters appear to be vain musings by an outdated luddite, I want to note my firm belief in the continued power of individual, exactly matching terms, for maximizing efficiency and impression share when historical performance on specific terms matches an account's objectives and budget (especially when limited).

Whether Google eventually kills the keyword outright remains to be seen, but for the time being, it certainly doesn't hurt to have an account focused on both (as I have argued previously in this book).

---

[7] Patrick Gilbert, *Join or Die: Digital Advertising in the Age of Automation*, (Maitland: Mill City Press, 2020).

[8] Chris Reeves, executive producer. "Evolution of the Keyword," PPC Ponderings Podcast, Season 1, Episode 1, October 2021, accessed 31 January, 2024, https://zatomarketing.com/blog/episode-1-ppc-pondering-podcast-evolution-of-the-keyword.

4 CONTROLLING THE KEYWORD FOR SMALL BUDGETS IS CRUCIAL

Most PPC advice given online revolves around accounts spending a great deal of money each month. How should you think about XYZ tactic when you have a gazillion dollars a month to spend on marketing, with a trillion for Google Ads alone? Tools, tips, and tactics abound for these accounts as they have the most important thing needed for good management, whether driven by Artificial or Human Intelligence... they have enough data.

But what about the little guy? What about the poor little account with not enough traffic, not enough conversions, and not enough money? Is it hopeless for that account? Ironically, these are who automation is supposed to help the most, when in reality it's who they help the least... since automation needs buckets of data in order to work well. I'm not saying you can never make Performance Max work for small accounts (we definitely have it humming in some of our MAKROZ SMB Accounts spending under $10,000 per month). But also, by Google's own admission, it is likely to not work well without enough conversions per month (how many conversions is enough? That answer ranges from 30-150 based on who you ask, but suffice it to say it's a lot more than the typical local business account will ever conjure up).

The thing with automation is that, especially in regard to acquisition and prospecting, it takes time and money in order to identify what works, what doesn't, and then what else works based upon what the system learned worked. It's really remarkable stuff, but the problem is that many smaller budgets don't have the ability to survive the learning phase of Google expansion efforts. Few small accounts have the budget ability to go broadly enough on expansion efforts for machine learning to work well.

Why not simply limit Google's expansion efforts with lower budgets? Well that's problematic, since that simply slows down the required learning period. You'll still need to expend the same budget to get the necessary data required, but it will just take longer now.

So what's the solution? I think it's the targeted, exactly matched keyword.

When considering how to utilize keywords (or really any targeting) in your Google Ads account, I see two things at play:

(1) the need to capture all the things (close variants and audiences are great for this!). I would lump this under expansion or acquisition or prospecting efforts. In this way, Google automation is fabulous. It allows for bidding signals such as landing page content and previous search behavior to identify target audiences in which to bid on, regardless of the specific keyword targeted. It's more about the "person" searching than their term. This category is where I think campaign types like Demand Gen and Performance Max fit in nicely.

But as we have discussed already, the problem remains: what if you don't have the data or the budget to accrue that data? This can result in the effort simply "not working".

"We tried Performance Max, but it didn't work."

Did it not work, or did it simply not fit within the required budgetary constraints and efficiency parameters? This is especially crucial to keep in mind for SMB accounts and is why I think there is a better way for beginning the SMB account in selecting your targeting method.

(2) the need to limit targeting to the most efficient keywords. Whereas there is a need for expansion in advertising, there is also a need for maximizing efficiency on core targeting methods. In Paid Search, that is (still) the keyword. I think keyword targeting is the way to do the second, especially in small accounts. But not simply in small accounts, as this can still be a legitimate strategy in larger

accounts. As we have already discussed in this book, since keywords reveal a user's spoken intent, they can thus be grouped into intent-based targets. Maximizing search impression share and click share on key terms that are a core part of a business's search presence are still a crucial way to efficiently invest ad dollars into the bottom funnel.

Google has always shined (until the last few years when close variant matching became more of a thing) in keeping targeted, specific keywords as a core method. I.e., "we are choosing to somehow limit our ad targeting to a select audience, in this case those people who type in this specific phrase we have found to be valuable based on past performance".

Consider, if you will, an expansion effort with automation that relies on Broad match or DSA rather than simply Exact match core keywords you know are your audience. In this instance, Google is identifying multiple potential opportunities that may hit your bidding targets, but then also invests a portion (I've heard it is likely around 20%) of your smart bidding into expansion opportunities. I.e., into untested keywords/audiences/markets in order to find new areas of potential revenue. That's a great option when you have the budget for it. Many SMB accounts that are getting into Paid Search don't need to know what can someday make them more money, but they need to know that Google can work for them now. Based on our previous discussion about needing enough time and/or budget, it's also unlikely that an expansion focused small budget account would even make it through a cash-poor business owner's patience cycle before finally giving up on whether Google will ever actually deliver on the "untapped potential!" promise of automation.

In that case, what they need is guaranteed intent, and that is most easily (and quickly) discovered by targeted keyword phrases that exactly match the business owner's target audience or offering. I've said this before, but I bid on terms for my wife's photography business here in Billings. We spend around $250-500 per month on Google, and almost predominantly those are on Exact match terms alone, along with some phrase match (and plenty of negative keywords). This is because we're not in a place where we want to dump money into "testing", but we just want to make sure anyone typing in "commercial headshots" or "newborn photography" sees my wife's ad and clicks on it!

This is why I think that Numbers 1 and 2 above should always remain a separate option within Google and specific keyword targeting should never disappear completely (regardless of things you may hear otherwise, such as "keywords are dead" statements!). In other words, I think Google should allow for semantic & audience expansion (in acquisition efforts), while keeping Exact exact for the limited budgets.

I think ideally a good PPC account of any size has targeted control over your core exact match keywords (since those are simply grouped audiences) and then utilizes Google's impressive acquisition options to expand beyond these core terms as profitability is demonstrated. Easy peasy – there is room for both in PPC.

I'll end by climbing on top of a soap box.

The desire to kill the primary targeting method ("keywords are dead!") within a semantically driven advertising channel makes no sense to me. I truly believe it would mean the death of Google's advertising business. They're not social media, and they're not primarily audience driven. Google is about words. Language. Communication.

If the core of what Google is about as a search engine is language, then it would be a mistake to move away from a language driven targeting as its primary targeting method. Google isn't about demographics, it is about words that people communicate. That is, has always been, and will always be the most powerful marketing targeting option for Google as a Search Engine.

Admittedly, Google's core business model has grown over the years to include apps, email, Display, and YouTube... but then those should be treated differently in a marketer's strategy, as well as kept separate in Google campaign types, bidding, and objectives.

The keyword may be killed someday, but that would be a mistake, since that would be the death of their medium: language.

If people "communicate" to Google what they are interested in, then it would make sense that targeting words is the best match. Early Google knew that, but will long-term Google survive the current revenue-obsessed mindset?

I guess time will tell.

5 TO BID OR NOT TO BID ON BRAND

Whatever your title in Paid Search Marketing, you likely have an opinion on brand term bidding for PPC keywords. What do I mean by "brand term bidding" here? It is simply the question of whether you should pay to show ads when a user searches specifically for your brand name in Google search.

There has been much discussion—and much written—on this topic recently. (Actually, there are 1.9M Google results for "should I bid on brand keywords in Google Ads" at the time of this writing). For those unfamiliar with the discussion, the reason for this particular debate is fairly simple:

Those without an investment in paid search tend to hesitate spending money on traffic that could have come for "free".

Those with an investment in paid search tend to prefer the inclusion of high-converting brand terms with high quality scores in their PPC accounts (and PPC reports).

There is a time to bid on brand keywords

So to be clear, I'm not a whistle-blower on paid search in this chapter by suggesting there are times when you should NOT bid on brand terms. I'm also not here to reveal a conspiracy that all paid search marketers have signed in blood to deceive the masses about the benefits of brand bidding for the sake of the industry.

In fact, I generally believe a brand should (nearly always) have a brand bidding strategy in their accounts. In 2024, the low cost of brand keywords, as well as the guarantee of your competitors appearing (whether accidentally or intentionally) in your brand auctions is pretty high. When a customer is convinced enough to buy from you that they type your brand into Google, the last thing you want to do is risk losing some of those to a cleverly written competitor ad simply because you didn't want to pony up the $0.43 for that click.

As some have called brand bidding, it's merely the "Google Tax" for doing business in PPC. Frustrating, annoying, infuriating, but necessary.

Perhaps you're yet unconvinced.

Here's a question for those who think you should never bid on brand terms:

If you don't specifically bid on brand keywords, do you then purposefully exclude all forms of brand keywords from all of your PPC campaigns?

If your answer is "no", then I'm confused. If you don't exclude all of your brand keywords from all your campaigns, then your ads will show for brand terms whether you want them to or not.

Example:
Let's say you (Polly) sell purple packs of pickled pumpernickel. You purposefully avoid bidding on any brand keywords, but you are bidding on phrase match or broad match "pickled pumpernickel" terms and maybe even exact match [pickled pumpernickel].

Here's the problem. If someone types in "polly's packs of pickled pumpernickel", which includes your brand name (on which you are NOT bidding), then your ads will still appear... on a branded keyword. With close variants these days, you'll be showing up for exact match search terms for your brand that aren't even close to as exact as the one above.

Oops.

So you are bidding on brand, just not intentionally. That's kinda even worse because you might as well break them into their own campaign to control your budget, bidding, efficiency targets, ad text, landing pages, and search impression share.

If your answer to my previous question is "yes, we are excluding brand keywords completely from our account", then I will first applaud you. Congratulations, you have followed your argument to its logical conclusion.

But after I heartily slap you on your back, I will have to slap you across the face. Nah, that's a little too violent; how about I just caution you to rethink your strategy.

Specifically, there are three glaring weaknesses with excluding all brand terms from your account:

1) If you added them in as [exact] match negatives alone, any brand term that a customer queries into the Googs will now be shuffled into your non-brand ad groups, and you will pay a far higher cost-per-click (CPC) for that non-branded query than you would have had you just bid on brand in the first place.

2) If you added them in as broad or phrase match negatives, you will actually be excluding your brand from any brand auction, and you wouldn't even know you missed it since it won't appear in your search query report. All those competitors bidding on "pickled pumpernickel" will have a heyday with those top three SERP slots on your branded query without any ads from you at all to protect your brand. Oops. Weakness No. 2 actually gets worse with e-commerce sites, thanks to the growing power of Google Shopping in the SERPs.

3) If you completely pull your brand from your account (that means Shopping Ads as well, along with Performance Max brand exclusions), then you are allowing far more advertisers to show in a highly clickable format above your organic results. Without Shopping Ad brand terms, there could be a dozen highly visual competitors' ads appearing above your first organic result.

(Side note: Regarding Google Shopping brand traffic, there is a brilliant strategy by which you can separate brand queries from non-brand queries in Google Standard Shopping campaigns so you can set unique bidding targets. Because this is such a popular strategy that still works, I have included it as an Appendix in this book.

There is a time to NOT bid on brand keywords

Charts, graphs, and arguments are great. In fact, there are many convincing charts, graphs, and arguments for why a PPCer should bid on brand terms.

But unsurprisingly, I've found SMB owners and general managers to be less-than-impressed with averages, should-be's and theories when it comes to paying for terms they could have gotten for free.

After all, when it comes right down to it, we are providing a service for real people with real businesses who feel the pain of every dollar in their marketing budget. This kind of emotional proximity can make things a little more difficult than simply showing a graph about incremental value.

Maybe I'm the only one who has ever received pushback from a SMB client when I pitch brand bidding, but I feel that I've only ever seen PPC articles that talk about brand bidding solely in a positive light. So my question for us PPCers now is this: Is it ALWAYS best to bid on brand keywords?

For instance, what about a brand search for a local coffee shop in which the Google places card displays with directions to your brand, your brand website shows at the top in organic results, your social media channels appear high up in organic search, a couple of news articles show up talking about your brand, and there are no competitor ads anywhere to be found. Is that a case where bidding on your brand really would lead to incremental traffic, or in this case, would you actually be paying money for what would have come for free? I personally think it's the second in this example.

After multiple conversations with concerned SMB clients, I have come to the opinion that it is not always best to bid on all brand keywords.

Based on this, I have three criteria for deciding when to avoid bidding on specific brand terms, and each case must match all of the following criteria:

Number 1 - The [brand keyword] is exact match - We discussed this briefly above, but if a keyword is a broad match or phrase, there is always a chance that someone will type in your brand, along with some random long-tail phrase you never considered but that a competitor is bidding on. In that case, you want to appear in that SERP so they don't win the day. Because of this regularly occurring scenario, I consider not bidding on exact match terms, but I typically find it's unwise to avoid bidding on phrase match brand terms since they can help you match to lesser-known auctions in which competitor ads abound.

Number 2 - The exact match [brand keyword] has no competition – On one hand, even if your brand dominates organic, those top competitor ads are tough to argue with if they're full, especially with mobile. Have you seen your mobile results lately for your brand? You might be horrified to see how much of that first tiny little "page" is taken up by "their" ads. On the other hand, if there is no ad competition, then the competition argument for bidding on that term loses ground fast.

By the way, just because you can't get a competitor ad to trigger for yourself doesn't mean they're not showing to other people. In Google Ads, a competitor could be excluding your location, have a capped budget during the time you are searching or be ineligible to show for your query for some other reason at the time and place you are searching. Because of this, we recommend running ads on all brand terms for a month or two, while monitoring your auction insights report for those exact match brand terms. If no competitors appear for those exact match terms in auction insights, you're safe to exclude… as long as the next criteria is also in place.

Number 3 - The exact match [brand keyword]'s organic results are dominated by the client's brand – Typically, a client's brand is likely going to be strong enough to dominate organic real estate on the first page of the SERPs, but some local businesses don't control it all (or some brands who have generic words in their name). This situation requires you to make a judgment call, but when in doubt, I'd suggest leaving your exact match brand terms live. As I said previously, risking a few bucks in the grand scheme of things to ensure you nail that high intent sale is likely worth it.

As a reminder, please note that ALL of these criteria must be met for you to exclude your brand terms.

Let's close by asking why go through the hassle? Why not just bid on all brand terms, collect the bump to the account's quality score, and move on with life?

It is because you are advertising for a real SMB company, run by real people who care about real money. I'm a big fan of the adage to "treat others like you want to be treated."[9]

If you are an agency, your client will be shocked that you are willing to lose conversions from the PPC pile for the betterment of the company as a whole. In reality, more accurate reporting helps everyone. Having a more realistic view of actual blended ROAS and performance from Google Ads without guaranteed sales that would have come in through organic (or other bottom funnel channels such as email) helps to set more realistic goals within Google Ads for both you and your client.

If you are in-house, your boss will be delighted that you are thinking of creative ways to maximize budget and conversions, therefore demonstrating that you also care about the company into which the business owner has poured his/her life.

This speaks volumes to SMB managers and owners. They have tight budgets and tighter fists. So if you present a strategy to help their brand be found without mindlessly bidding on every brand term that comes along, they will love you for it.

But also, I kinda think it's just the right thing to do.

---

[9] Popularized by a lesser known historical figure named... oh I don't know... JESUS.

6 INVEST IN MORE CHANNELS THAN PPC OR DIE

When we receive leads for new clients, our filtering system includes asking how much of their marketing budget is devoted to Paid Search marketing (the only channel we manage at my PPC agency).

Why in the world would a person who has invested his life in the realm of Google Ads purposefully shy away from taking on clients who spend all their marketing budget in PPC? Wouldn't we want to be the sole keeper of the keys? When a prospect proudly states "yeah, we actually invest about 95 percent of our entire marketing budget solely into Google Ads", why wouldn't I hear "job security" rather than what typically goes through my mind: "warning, warning, warning"? My primary motivation for being alarmed by a client's declaration of dependence upon the marketing channel to which I have devoted my life is five-fold. We will spend this chapter on each of those points. Here is why you should be alarmed when someone's business rests only in PPC.

PPC Can't (Really, Technically, Kind of) Create Demand

For the sake of oversimplification, let's assume there are two core sides to advertising:

- Demand creation
- Demand capture

PPC marketing is stellar at demand capture. As we discussed in a separate chapter, one of the things I love about paid search (I'm here using PPC and paid search interchangeably, just because I can) is that you nail the what, where, and when of an inquiry down to the individual level. You can get your ad in front of one specific person (amazing!) just by identifying what they are asking Google. You get it right in front of them when they are asking and marketing shines its brightest.

However, that is when they are asking about it. They had some prior knowledge of something in order to spawn a question or elicit a search. We don't inquire about what we don't know. Because of this, you need more than PPC in your repertoire to actually build and grow a business long term. That is, if you want to create demand for your product, you need to get in front of those people who aren't even asking questions about your product yet. This is the magical part of marketing: demand creation! And this is what paid search struggles to do successfully.

Here's where I hear a chorus of well-informed, intelligent voices complaining about my oversimplification. There are ways in which you can (arguably) use paid search to generate demand, but I would note that this is rare (and not typically a cost-effective way to generate demand), since keywords tend to get more expensive as time goes on. In other words, you may show ads for your "Massage and Yoga Parlor" to people inquiring on Google: "what are ways I can decrease stress in my life?" But, is that truly generating demand or simply positioning your brand to meet the demand already present? (I would argue the second.)

Even if you argue this to be demand creation, you can probably take 100 clicks for that keyword and get in front of thousands of your target audience on Instagram, YouTube, or Facebook instead. Thus to my point on cost-effectiveness. All that to say, if you genuinely want to grow a brand and generate demand (and not simply survive, maybe), you will want to invest in more than PPC, since PPC is primarily geared towards demand capture.

PPC Isn't Marketing, PPC Is a Channel Within Advertising, Within Marketing

I have encountered people over the years who understand paid search advertising, but they aren't good at marketing. In other words, they know how to set bids and choose keywords, but they don't know how:
- o You actually find people (or create an interest in people!).
- o Position a product well to them.
- o Convince them (usually over time) to invest in a brand.

As I have tried to communicate in this entire book, paid search is a fantastic advertising channel in which to invest. It does a great job of answering questions people are already asking and offering the solution of your product or service. However, if this is your entire marketing strategy, then eventually it will come to bite you simply because you need some way to create demand in the first place.

That is, you need some way to:
- o Continually engage with people (social media).
- o Drip great content over time to remind people of your value (email).
- o Invest in what you can grow over time without shelling out cash for every dad-gum click (organic).

And those are just digital channels! If PPC is the only place you have invested time and money, that will eventually come back to bite you. Find a PPC agency who will care about your business success, not just be willing to ride a quick wave spending as much as they can to capture only the low-hanging "demand capture" fruit before heading out the door.

Let's really build a brand! And to build a brand we need more than PPC.

PPC Can Change (& Has Changed) at the Drop of a Hat

This point can't be understated. You could invest years into an advanced structure in Google Ads that works exceptionally well, only to watch Google shut it down by eliminating some functionality on which you were dependent.

A few years back, I was assisting an agency with one of their clients, and we were seeing exceptional success by segmenting our search campaigns out by device. Then Enhanced Campaigns hit and Google removed our ability to segment by device type (they eventually gave device targeting back to advertisers) and this particular client took a beating in their AdWords account. Sadly, it wasn't only the client who "took a beating", as they also took it out on the agency with stress-filled calls and hand-wringing—all because Google made one change.

It. Can. Happen. To. You.

For the love of all that is good, don't put all your eggs into one basket, diversify your marketing and survive the fickle nature of channel changes! I'm going to speak to agencies directly in these next two points. The previous points can be helpful for in-house managers, or executives to consider as well as agencies. These next points fit directly into the unique agency/client relationship.

For Agencies: The Stakes Become Outrageously High for Every Test, Experiment, Bid Adjustment or Disapproved Ad

I invest myself emotionally in each of our clients. I feel wins and losses in our client accounts, and I believe my team does as well. However, I have found that, for clients who primarily invest all of their budget into PPC, the emotional toll of high stakes is even higher than normal and typically not worth it as an agency.

When a client's entire marketing plan is PPC (a role it was never meant to have), it allows for a variety of profit-killing behaviors to occur, such as: Obsessive, micro-management, communication regarding intricacies in the account. "I noticed the bounce rate on this keyword is 5 percent higher. Why do you think that is?" "Can we have a call?" over every minor dip or bump that happens in an account. "Should we pause this ad you pushed live yesterday?? Did that cause this morning's dip?!" There is an increased willingness to fire the PPC agency with the belief that a change will fix things, or something, hopefully, doggone it something, please work, ANYONE...

I have found this sad truth to be the case: the stakes for your agency and personal emotional well-being are substantially higher when a client isn't diversified in marketing. And I don't mean that in an applaudable, "you got this, buddy, go get em. Reach for the stars!" encouraging way, as if you just need to toughen it up and meet the challenge. I mean it like, "the stakes are so high because they don't understand marketing and are putting unnecessary pressure on you because their business literally rests on your shoulders." I personally don't want to be in that position because of the previous three points. It's not my responsibility, and it can't be because it's asking PPC to do something it cannot do. Unless something significant changes in that business, it won't last much longer anyway (or at least, your relationship with them won't last past the next big Google change #realtalk).

A Client Who Cares Solely About One Channel Is Probably Difficult to Work with Because They Do Not Understand How Marketing Truly Works and What Is Required to Actually Build a Brand

Lastly, we come to a difficult, blunt point that I hope isn't taken as harshly as it may sound at first blush. We are all learning and growing in our understanding of how this all works. There may be a legitimate time early on in a company's life when someone admits "here's the position my business is in with marketing. I don't want it to remain like this, but we need to emphasize PPC right now. Please help us diversify and here is how we are intentionally invested in doing that."

Sometimes startups have a lot of PPC investment to begin with because they're trying to rapidly ramp up while they work hard on building out their other channels. That's a great way to start a business, and it demonstrates that someone understands marketing because they are using one channel to build the others. Of course, when it comes to startups, recognize that they have their own challenges in PPC, so make sure to walk in with eyes wide open and an extra budget for additional communication. Some startups want you to act as their CMO, not their PPC agency, so be especially cautious of scope creep in your agreement where you end up doing more work than you originally agreed to do.

On the whole, however, I have found that established businesses with a significant amount of their budget invested in PPC have purposefully built a business that way. They've leaned too far into PPC, and it will come back to bite them eventually. For example, I have observed a certain type of company culture that simply does not understand the first few points I brought up above. Those executives don't want to hear what you have to say regarding Top of Funnel, demand generation, brand positioning, etc… they just "want PPC results!" "Yeah we get it, whatever, that's not your concern… but seriously can you 3x our revenue and hit ROAS goals next week?"

If that's the case, my unsolicited advice is to pass on those clients (we'll discuss later how to better identify and avoid bad clients). They will work for a while, but the first issue that occurs in the account will have them shopping around again for another PPC agency who they're hoping "will be the one." They don't understand marketing, and thus there is a core problem within their business that you, the PPC agency, aren't going to fix even if you hit their ROAS goals. You've wasted significant time onboarding, communicating, and setting up a client who isn't a long-term, viable option. You may find it worth your time to refer them elsewhere right from the beginning.

So, did you ever think you would hear a PPC agency concerned about working with someone who has 90 percent of their marketing budget in PPC? You have now! I hope you take those five points to heart. May they help you be more cautious the next time you meet someone who has invested their entire marketing program into PPC.

7 WHEN TO BLAME THE POOR POSITIONING, NOT THE PPC TARGETING

"We'd like to bring you on because you have specific experience in this unique apparel niche demographic."

I was flattered, as well as a little uneasy. Sometimes the most difficult people to please are those who already hold you on a pedestal. Nowhere but down, and all that.

"We do have experience and would love to invest energy into getting your ads in front of that specific niche demographic."

Fast forward 14 days. We had overhauled their Search and PMax campaigns, as well as Product Feed. I was pretty excited. We weren't always able to push out that amount of change fast, but we had the bandwidth and we were close enough to Black Friday that I wanted to make sure we wrapped up new campaigns early enough to stabilize before the busy holiday season hit.

Fast forward another 14 days. "We're going to have to let you go, as we haven't seen the performance increase we were expecting when we hired you."

I blinked. I checked reports.

We had increased, this is not an exaggeration, their traffic somewhere around 1000% for the ultra-niche demographic we were hired to better target. They had a unique group of particular people they wanted to get in front of. People searching for a specific problem that their apparel specifically solved. We increased exposure on just those terms (through optimizing Product Titles and other Feed Attributes, and making campaign targeting adjustments) and drove traffic at a rate of thousands of percentage points higher than what they had been getting before. I mean, they weren't getting much traffic for those terms, so we simply turned on the floodgates.

On the flip side, he was right. The traffic, who was completely dialed in based on how they were searching, wasn't purchasing. I pointed that out and made suggestions about how their product was being positioned on their website and landing page.

Nothing changed, and we got fired after 30 days of management... having done, in my opinion, exactly what we were hired to do. Look, the point of this story isn't just to garner sympathy (okay, maybe a little... cause that sucked). The point of the story is to note that product positioning and ad targeting are often conflated, but they are two very different things. They are also controlled, typically, but two different groups of people. In this instance, the right audience was being sent to the right product, but they weren't actually purchasing the product. In that case, finding a different audience isn't the solution, since the current audience is already ideal. The question must be asked, "why are the people who should want to buy my product not buying my product?" Perhaps it is because they have not yet been sold on why they should buy your product.

A common trap I see brands fall into with advertising on Google Ads is to label a positioning problem as a targeting problem.

Why aren't people buying on this term that is perfectly in line with your product? It's not because you haven't yet figured out how to best segment out the ideal Google Search Ads audience within an audience (though certainly there can be times when a change in ad targeting is required), it's because your brand hasn't figured out how to position itself well in front of your ideal potential target audience (shout-out here to go buy and read April Dunford's excellent book <u>Obviously Awesome</u>).[10]

Let's say your store sells cardboard boxes for shipping. But not just any cardboard boxes, you sell high quality, ultra-durable, cardboard boxes for shipping. The thing is these tend to be 45% higher in cost.

You look into your Google Ads account and notice that you get a lot of traffic from people who search specifically for "best cardboard boxes for shipping", but your CVR is really low. This means your ROAS target is well below your overall goals. Your boss tells you to trim some fat in the account, and so you shrug and exclude [best cardboard boxes for shipping]. I'm not going to lie, this decision-making process kills me.

Before overhauling your ad account, changing up your target audience, or firing your ad agency, you need to pause, take a deep breath, and ask yourself why it is that people searching for exactly what you are selling aren't buying, and then you need to figure out how to sell to the audience you are already successfully targeting. You have their attention, and perhaps even their interest, but you don't yet have their wallets.

How can you determine this? Some thoughts on how to gather this research:

- surveys (Why are people buying competitor boxes? Ask people that question and listen to what they say.)
- industry forums (Who is sharing the dirt in some reddit community about shipping box purchasing?)
- customer reviews (Why did some people decide to buy yours? What do existing customers like about it?)
- competitor research (Read your competitors' reviews! Why are people buying their boxes and not yours?)

Then, leverage that into better positioning your product/brand to address core reasons a person wouldn't normally want to pay more for shipping boxes.

Perhaps you create some User Generated Content (UGC) of customers who talk about how much they wasted in damaged shipping costs until they switched to yours.

Perhaps you add an actual slider on your landing page of how

---

[10] April Dunford, *Obviously Awesome: How to Nail Product Positioning so Customers Get it, Buy it, Love it*, (Ambient Press, 2019).

much money damaged shipping costs are to a business and how they actually save money by paying more for your boxes and less for fewer returns/damages.

Perhaps you have an even better idea since you know your business better than I do?

My key point here is that advertising and marketing can only solve so much for you (as you can read in another chapter in this book as well).

Marketers have limitations.

Marketers can identify and send the right audience, but we can't actually force them to buy anything (legally or ethically), and it's up to you (or your brand owner client) to ponder *why* the core audience isn't purchasing.

Do you have a positioning problem? The easiest way to tell that you have one is if your targeting is dialed in and yet people aren't purchasing.

Well, or your cart checkout page is down. But that has an easier solution.

8 THE MARKETING FUNNEL IS NOT DEAD

Declarations of people dying who aren't actually dead isn't just a phenomena of the Social Media era. In 1897, people apparently didn't pay much attention to detail; either that, or Mark Twain and his cousin, James, looked an awful lot like each other. When James passed away and people mistakenly thought it was actually Samuel Clemens (Mark Twain's real name), he had to wire from London the now famous phrase, "reports of my death have been greatly exaggerated."

People have always had a fascination with death, and they love to kill something off. There's a reason only a couple of people (if that) make it through every horror movie. Marketing is no different. People love to kill off non-dead things in PPC as well, believe it or not. One of the easiest targets is picked on so often simply because it is so significant: the marketing funnel.

But is the marketing funnel actually a dead concept? Well, in the words of Mark Twain, the reports of the funnel's death have been greatly exaggerated.

Why kill the marketing funnel?

Unless you've been living under a rock (stone/pebble/hard surface… that's a keyword close variant joke, sorry not sorry) in the digital marketing world, you've likely run across at least one confident declaration in the past 12 months that the traditional

marketing funnel concept is dead.
Or dying.
Or maybe maimed.
Or at least, shot in the foot.

The marketing funnel, for those unaware, is a concept created in 1898 to describe the "theoretical customer journey" and assist in targeting the right message to the right person at the right time. If, for instance, someone is just curious about your business, you don't want to "ask for their hand in marriage" by pushing them too hard for the sale, but rather "court them" by focusing on answering their questions and moving them to the next stage of the funnel. The funnel is typically seen as four categories (see below) that move from a broad group of people (cold leads) to a smaller, more targeted, and closer-to-purchase group of people (hot leads).

Those four categories are also known as AIDA:
- o Awareness
- o Interest
- o Desire
- o Action

The logic of AIDA is that people who have a desire for your product are fewer and closer to purchase (from you or a competitor) than people who are just becoming aware of your product. Thus, you need to cast a wider net at the beginning to get in front of people and then move people further "down the funnel" to get sales.
Easy-peasy, right? Well, sort of.
The marketing/sales funnel had been somewhat widely accepted until we 21st Century chronological snobbists began to scoff it away and proclaim that digital changed everything inexorably and irrevocably. But has it? In this chapter, I want to examine whether the marketing funnel really is outdated and kaput because of the influence of digital.
I then hope to prove that, in fact, our understanding of the marketing funnel has simply evolved and we see its intricacies more clearly than ever before... while the core philosophy remains the solid foundation holding it all up. I think this is important to do, by the way, because I find I use the marketing funnel constantly in

PPC Management. When I am considering building new campaigns, for example, how I think about the audience, bidding targets, campaign objectives, and ad creative/text all fits into where I see this campaign fitting into the broader customer journey. This is not a philosophically pointless argument. I think it's important for us to grasp as Google Ads practitioners!

So I realize there is a lot of philosophy here (then why did you pick up this book, if that annoys you?), but I think that is core to providing a good foundation for our tactical advertising choices... so let's dig into the arguments.

Legitimate Concerns with the Marketing Funnel

Let's hear from the marketing funnel haters first. But perhaps that's not entirely fair, as there are those who question the validity of the traditional funnel for good reason. The general consensus of the "funnel haters" I've run across is actually quite legitimate. That is, they posit that user engagement with a business is far more complex now than it was in 1898.

This is true, and an important note, though I don't think it undermines the funnel concept. How? Here is how the anti-funnelist typically makes his case: "People can interact with a business in multiple ways on multiple devices (many of those completely untraceable visits), so we need to think of a new way to describe this besides the marketing funnel because obviously the user journey is more complex than it used to be."

What I take issue with here is the assumption that the user journey is more complex than it used to be. I would ask in response to the previous charge: "is the user journey truly more complex in the broader customer journey categories themselves, or are we actually more clearly seeing the inner workings of each broad category?"

That is, should we rethink the structure itself, or is what we are seeing in its complexity simply a more advanced look into the mechanism of the funnel itself? I would suggest that nothing has changed in these primary principles of the marketing funnel (i.e., the structure itself hasn't actually changed) and it is, in fact, the second option in my statement above.

Example:
Think of it this way. You buy a house at 1598 Penn Station that is a complete dump. Your plan is to completely overhaul every square inch of the home, to update all of the furniture, siding, windows and doors, and to make this a new home in every way. But is it a new home, or is it simply an updated structure? In that way, the home is still the same home at 1598 Penn Station. You have just updated the existing structure and even caused it to live up to its full potential. But it's still the same house.

In this way, I would argue that the overall AIDA structure is solid (even if the windows need to be updated, continuing with our metaphor).

In Defense of the Marketing Funnel

So what is the structure of AIDA that remains unchanging as we consider PPC marketing in 2020 and beyond?

Here are the two core, unchanging principles I see in the marketing funnel:
1. There are stages of customer intent that change over time moving from less to more intentional in purchase behavior.
2. These stages of customer purchase intent tend to correlate with the number of people in those stages (i.e., more people in awareness, fewer closer to purchase action).

In this way, I think the best way of illustrating marketing today is an adaptation of a HeroConf presentation by my brilliant friend Aaron Levy of Tinuiti.[11] Levy described the concept as a "maze", with the helpful inclusion of the "Top/Middle/Bottom" qualitative intent language to describe some sort of inter-categorical

---

[11] Aaron Levy. (2017) "Transforming Multichannel Marketing Into Omni-Channel," Slideshare, accessed 31 January, 2024, https://www.slideshare.net/hanapinmarketing/transforming-multichannel-marketing-into-omnichannel.

movement (which is, if you remember, the first of my core unchanging principles of the marketing funnel). In other words, while there is much mess happening within the AIDA funnel, there is still the overall actual structure that continues... Top to Bottom of funnel traffic.

But isn't the world of Digital Marketing so much more complex than AIDA? Sure, we have social which is new from 1898, and email, and paid search, and organic, and in-store visits, and Amazon store visits and all the rest of the things that come with marketing today. But overall, it is my belief that these visits all still arguably follow those two core principles of the marketing funnel.

Okay, so what about PPC? This chapter is exceptionally relevant to PPC. At ZATO, we constantly think in terms of the funnel as we consider what Search Terms or Audiences to target and what content to show them based on where they are in the Funnel. It informs our bidding decisions (we're going to bid more aggressively for people closer to purchase), and it informs our campaign structure (we try to keep audiences or keywords grouped according to where they are in the funnel for bidding and targeting and budgeting purposes). With PPC, you can still identify unique audiences and where they are in the funnel, and then craft a message that meets them within that funnel.

In Conclusion

If you don't see any correlation to the marketing funnel in your marketing, perhaps the weakness isn't in the funnel itself, but in your business audience identification capabilities. Improve your ability to better target your ads to your audience by considering how your targeting fits into the four broader AIDA funnel categories and then craft messaging/creative to support your targeting. As we consider the concept of the marketing funnel, it's tempting for us, in the midst of the maze (mess) we marketers try to track every day, to throw our hands up and exclaim "it's all chaos!" However, I have argued that there is still a broadly identifiable purchase intent progression that takes place in the customer journey. That is, and always has been, the AIDA marketing funnel. Rather than kill it, we need to realize the genius of it. Even 120 years later, it still enhances our ability to understand the different phases of the buying journey.

Even if we have more access into the true mess that is each "phase"... and even if those phases sometimes do shift around or get skipped in the messiness of actual consumer purchase behavior, the overall customer funnel structure remains.

Long live AIDA!

9 WHEN OPTIMIZING YOUR ADS STRANGLES THE FUNNEL

Perhaps you've heard The Byrds croon out these famous lyrics:

To everything (turn, turn, turn)
There is a season (turn, turn, turn)
And a time to every purpose, under heaven
A time to be born, a time to die
A time to plant, a time to reap
A time to kill, a time to heal
A time to laugh, a time to weep[12]

In these words,[13] we're reminded that specific actions often have specific purposes… it's just all about timing.

As we think about PPC, I can't think of a more important principle for auditing low conversion rates in your account than this. The reason this principle is important is because there is a strong temptation for a client who has hired an agency (or a boss and a PPC employee) to become overly fixated on repairing the marketing, when in reality the marketing may not be the problem.

---

[12] Pete Seeger. "Turn! Turn! Turn!". *Turn! Turn! Turn!* Columbia Records, 1965. Vinyl Record.

[13] They originally appear in the Biblical book of Ecclesiastes, by the way!

It takes a deep knowledge of the account and advertising space (as well as a little guts) to push back on this, but pushing back may be the very thing that saves the business. If you focus on filtering audiences, or some other marketing "button" to push, when that isn't the necessary solution, you will strangle your marketing funnel and the account will suffer a slow, painful death.

As a note, I realize we already discussed this in a previous chapter, but I think it's so important I wanted to bring it up again, framed differently (specifically this time around the customer purchase funnel).

Let me illustrate what I mean practically before offering a solution.

A Painful Illustration To Which You Can Probably Relate

Your client is a lawyer in California who wants to offer a solution for people whose insurance companies screwed them over during wildfire season. You are the PPC analyst, and you identify and bid on awesome audiences targeting people who are actively interested in insurance claims, as well as generally target people in specific geographical areas impacted directly by the devastating wildfires. You also dig into Search queries and identify high intent terms to target.

You have ultra-targeted Exact match terms in the account, and even some Phrase match terms to pick up other targeted queries with the budget you have left. You have checked Search Query reports and can vouch for the targeted queries sending traffic. In fact, you're killing it with high CTR and average positions on those top terms and you have to admit that you're a little proud of the account. I mean, you nailed it.

The campaigns are pushed live for a couple of months, contact forms begin to be filled out and customer info is sent left and right (picture here George from "Seinfeld", snapping his fingers). You optimize your bids and ads based on this, and everything is just humming. What a great account!

You smile.

Then the client calls and your smile fades as the call progresses. "The campaigns aren't working! What can be done?" They begin to micromanage your every move. Is this keyword problematic? Why did you use "the" instead of "a" in the ad? Those Facebook ad pictures should have featured a brunette instead of a blonde model. They keep asking questions about, you know, the really important details you hadn't yet thought of that will really move the needle </sarcasm>.

You learn through this process that the contact forms aren't converting over to actual cases for the lawyer, and the client is convinced your targeting is the reason. What do you do? Unfortunately, I have described a classic, sign-up to subscription drop off (in sales language, MQLs that never become SQLs, or even SQLs that never turned into paying customers).

So what do you do when lead quality dips? There are two primary places you can look:
- Audience (the marketing)
- Post-Lead Optimization (the client)

Let's consider each of these as we determine the best way forward.

Audience Optimization (your marketing)

In the above issue, your client is naturally going to want to focus on audience filtering because, frankly, it's easy for them to do so. It's someone else messing up the work. It's the third party who must have screwed something up, and frankly, there are a lot of third parties who make terrible audience decisions so they're not wrong in looking here.

In this step, the focus is on the type of traffic being sent. Can we improve our ad timing, ad creative, keywords, bids, devices, or a host of other targeting settings? This can be tricky, of course, because there is nearly always improvement that can be done in marketing. However, while this needs to be investigated (and if you are a good PPCer worth your salt, you are already digging into this), this is certainly not the whole story, and there is a real danger (perhaps you are now thinking of an exact conversation you've had in the past) of becoming so fixated on the marketing aspect that the client-side responsibility is neglected.

Post-Lead Optimization (the client's responsibility)

This is the second aspect of improving lead quality and one that resides entirely on the client side (in most cases). It takes a good deal of work and requires a good deal of buy-in (and humility) from the client. In this, the argument is made that the audience is not the primary problem but rather the client's ability to convert the audience being sent.

Go back to the lawyer illustration from before. In this instance, the client may attack the keywords you have selected by saying something like: "People who type in attorney rather than lawyer have shown us to be a less valuable audience. They have a higher lead to case-close ratio, so please stop bidding on the word 'attorney' and we'll see that CPA lower and be happy." This makes a lot of sense (and it actually could lead to a lower directly tracked Cost per Lead), but there is a giant, gaping, account murdering, monster-sized hole in this logic. Can you spot it?

Strangling the Funnel

The giant gaping hole is that the client is asking PPC to do that which it cannot do: convince the user of the value of their services... and in doing so, the request to limit the (good) audience to eke out a better conversion rate is actually killing what marketing is supposed to do: fill the funnel.

Fill. The. Funnel.

Yes, the funnel needs to be filled with a certain level of quality audience, and we can't just find cheap clicks from bots in Outer Mongolia in order to fill the funnel. But I know that, and you know that. So let's assume for the sake of this logical progression that we are all good marketers sending a qualified audience to our client. If that is truly the case, then it's now up to the client to sell that solid audience on their value.

I'll say it a different way: marketers send qualified audiences, but it's up to the website (client) to convince that traffic to convert to the macro-conversion (the main action you are trying to get your target audience to take).

What I mean is, if we filter our target audience by preventing certain kinds of people from actually completing the form submission (or even seeing our ads), then we can likely increase the Lead to Purchase ratio... but we also run the risk of not ultimately filling the funnel with potential customers who should have been willing to convert. It's a ditch to be avoided on either side of the road (I like that analogy, don't I?). We must avoid steering into the ditch of an audience too broad to be interested in your client's product, but also avoid overcorrecting into the opposing ditch by failing to recognize and convert cold audiences who may help grow your brand down the road.

Have you noticed that some brands really just explode with growth (I mean here, the good kind of profitable growth, not a scaling built on a house of cards where they are simply outspending their revenue)? They seem to grow faster than they should, they outpace others, and they just have it figured out. How do they do that?

An account will ultimately see exponential growth, **not by limiting its customer base by continually filtering out people actually interested in the product,** *but by successfully identifying how to convert more of those not-yet purchasers who are already showing interest.*

How do you identify when it's not an audience targeting issue? The big "tell" is when a portion of those who have expressed interest in your product/service (who may even move to submit the macro-conversion) are then not ultimately convinced by the site or follow-up process as to why they would get increased value paying for your product/service rather than going off to find another option.

That's a huge red flag. You mean to tell me that someone made it all the way to signing up for your free demo, but then you couldn't convince them to purchase your subscription? Doesn't sound like a targeting issue to me. That's a person who was interested enough to be interested but not enough to pay because you (or your client) failed to show them your value. Don't fix the marketing there; fix the product or offer or content or pricing.

When to Fix Audiences & Change Your Offer

Now, beware of something that happens at this stage in the game, which is the temptation to turn to Conversion Rate Optimization as the solution. "We are only converting some of the people who sign up for a free demo, so maybe… let's change the Purchase Subscription button to the color green!" Your very expensive, very young business consultant tells you, "Green means go! Yes!"

This is where identifying the right kind of conversion rate optimization is hugely important. Conversion rate optimization, with the intent of shrinking the funnel, can be an account killer. If you think the way to solve your lead problems is by limiting a good audience, then you'll never blow up (in a good way). This is because you're actually shrinking a solid audience, an audience who is actively interested in what you have to offer enough to sign up for your first step. Now you still need to convince them of the reason they should choose you over your competition.

It doesn't mean they should be bounced from the website so your Conversion Rate numbers look better in your monthly report; it means you need to sell to them!

That's not marketing's fault, and it will never be. Marketing is supposed to fill the funnel with great traffic. The website's job is to convince that traffic why you're the right choice.

So, to summarize: are you sending people through you advertising who are disinterested in your product because of your poorly chosen audiences or keywords? Then fix the marketing problem.

Or, are you sending people who are searching for the right queries and taking the right micro-actions on your site, but who do not yet want to purchase? Then fix the core business (not a marketing) problem.

Fix your offer, step up your email strategy, optimize your

landing page speed, change your pricing, or even rethink your business model. I love marketing and think it's amazingly powerful, but the best marketing won't fix a core business problem, no matter how much you plead with it.

10 PPC ISN'T PRIMARILY FOR SCALING

Maybe the most difficult thing to admit for a PPCer is that rarely (never?) do great PPC tactics themselves scale a business... though poor PPC tactics can absolutely prevent a business from scaling (or scaling as fast as it could).

So what DOES actually scale a business then?

Hey, that's a great question! You're pretty smart.
What actually scales a business are things like:
- fantastic creative/video
- a Product Detail Page (PDP) that actually sells to the potential customer landing on the site
- the right product positioning
- the right product pricing
- smartly done promotions
- well-planned & executed inventory management
- wise cash management (especially around budgeting to the right channels)
- Etc.

See a common theme there?
All of those levers are owned by the brand. In this case, the brand operator is the super hero here, not the agency! But that is

often flipped around. The operator seeks out a Google Ads agency hoping they can work super hero level miracles. It's as if Clark Kent keeps trying to find someone who can help him, instead of just ditching the glasses (I mean, are those really much of a disguise??), rolling up his sleeves, and getting busy doing the thing he is good at.

The brand operator has far more ability to scale a brand with good decision making and execution than does a Google Ads agency. I think more businesses would be more successful with Google Ads if they stopped trying to scale the channel itself (and thus spent endless time and energy and money "tweaking things" or "finding the perfect PPCer") and spent more energy, time and dollars on solving the above problems.

I know that may sound weird coming from a PPCer, but I dislike being expected to solve a problem I can't solve, so I think it's a worthy topic to include as a short chapter in this book.

For you PPCers reading this, perhaps this means you become great at sniffing out when a brand operator is flipping this around on you. Are you expected to deliver something you are unable to deliver? The best operators respect being tactfully told what is best for their business, and sometimes the best thing for their business (your client) may be to simply point out the obvious: your campaigns have limitations, they are already targeting the right audience, and perhaps there is a bigger problem at play here that needs to be addressed from the brand side? If you become skilled at pointing the brand operator to those potential triggers I listed above, they might just treat you like a superhero.

By the way, to be clear,
- o I absolutely agree that a poor PPCer can harm a brand with bad tactics and account optimization decisions.
- o I absolutely agree that a poor PPCer can hinder a brand from scaling as rapidly as it could (if all the right levers I noted earlier are set) with bad tactics and account optimization decisions.

So, I'm not saying there is no impact to a brand between a good or bad PPCer. The brand still needs someone who understands the best way to get the most efficient growth possible in a Google Ads account... but I maintain that the primary scaling levers are within the grasp of the brand operators, not the PPC operators.

11 BALANCING GROWTH WITH PROFIT OVER TIME: THE ASCENDING SEESAW OF SCALING

"We want you to 6x our revenue at the same ROAS by next month. We have lots of opportunity in our space!"

Have you ever heard this request from a client (or prospective client)? It can be very common, especially in spaces where a smaller startup company sees an opportunity to disrupt a major industry. You want to "disrupt" the shaving industry, which impacts nearly every adult in the US? Then you're correct in identifying opportunity; it's simply the expectations that need to be reanalyzed.

The purpose of this chapter is to detail out how I (and incidentally, ZATO, my agency) view PPC spend growth. If you're tempted to blow past this chapter and yet are simultaneously tempted by the growth belief system voiced at the beginning of this article, I would encourage you to read on, as this is crucial to understand for long-term success.

The Ascending Seesaw of Scaling Awesomeness (or ASSA)

The following "ascending seesaw of scaling" metaphor is how I have historically described growth to clients, and I'd like to share it more broadly. To my knowledge, there is no better metaphor or word picture, though I would be delighted if you would share yours with me on LinkedIn or Twitter (I am @PPCKirk; let's connect while we're at it)!

There are Two Key Aspects occurring in PPC account growth for which an "ascending seesaw" is the best mental image I can conjure. Those Two Key Aspects are: ROAS and Revenue are conflicting PPC goals in rapid growth (seesaw). Your business should grow in top-line revenue over time (ascension).

Let's dig into each of these.

(1) ROAS and Revenue Growth are initially conflicting PPC goals in rapid growth.

First, let's look at the idea of overall PPC goal setting and how that impacts account optimizations. When we ask a client what their goals are in an account, we try to frame it in a way that tells us whether they want to grow their revenue rapidly or to focus more on current market share ownership and maintaining profitability. This is crucial to understand as a PPC account manager because it completely changes everything in the account.

- What types of keywords and audiences are you going to target?
- How aggressively are you going to bid?
- How are you going to determine what devices, regions, and keywords to pull back bidding on?

All of those questions and more come directly from this primary question of what your overall business goal is at this time for the PPC account. The reason ROAS and Revenue Growth are initially conflicting goals in an account is because one is focused on saving money, and the other is focused on spending money (albeit, wisely).

Practically speaking, when we push harder for rapid growth that typically means: bidding higher, finding new keywords and audiences to target, and shifting strategies to enter upper funnel auctions (which will tend to have lower tracked ROAS).

All of those have the practical result of surging spend and traffic from previously untapped auctions. Remember, Google Ads is an auction based system where you bid for position on every auction. When you bid higher, you can actually enter new auctions (or in some cases, simply bid more for the same auction and click).

When you target new keywords or audiences (whether upper or lower funnel), you are entering new auctions. This means you need time + money to make optimizations to return profitability to desired numbers over time as you learn which auctions are worth continuing to pursue (they are profitable), and which should be ignored (unprofitable).

This is why it's a seesaw. As ROAS grows, revenue will shift lower. As revenue grows, ROAS will tend to shift lower. Back and forth, back and forth... (until your 4-year-old jumps off at the bottom and plummets your 2-year-old to the ground screaming... but maybe my kids are the only ones who do that).

Example:
Let's say you are a company selling high-end $70 razors that has seen a lot of success in bidding on tight, upper funnel keywords such as [luxury razors], [high end razors], [best razor under $100]. You are pretty darn excited about this, but you have noticed that your traffic is fairly limited. You own the tightly controlled terms but want MORE of the market.

You enlist your PPC agency to begin targeting more upper funnel terms such as [best razors] and [razors for men] and [razors for women]. You tell them you want to spend DOUBLE (WOW, WE ARE SERIOUS NOW) and see what happens. You're stymied and shocked when you see ROAS drop hard next month, even though you're now spending double. You ask the PPC agency to fix the ROAS, and they tell you they are working hard on optimizing the new keywords, but need more time.

You roll your eyes, TYPICAL PPC RESPONSE, and fire your agency.

The next five agencies can't meet your expectations either, and you start to tell people that PPC agencies are shady (I mean, in your defense, there are a fair share of shady PPC agencies).

What happened?

Well, getting super practical for illustration purposes, you doubled your ad cost by entering new auctions for these keywords, but in this instance two specific things happened: Your new keywords needed more time to gather information to actually make the best decisions. You entered more upper funnel targeting and now need to adjust ROAS expectations to match (Biz 101: you won't make the same profit on a person entering your sales funnel as you do on a person at the end of the funnel, but that's because the person at the end of the funnel also spent more of your money elsewhere making their way through your danged funnel in the first place). By the way, that's also why Brand and remarketing campaigns should actually be required to have a higher ROAS, since they've already cost you earlier in the funnel.

Anyway, your additional budget got sapped up on the new keywords, but when they actually got in to investigate, your agency could see they didn't have enough spent on individual keywords yet.

This means they just did not yet have enough individual data to make a great decision about whether these terms were actually winning over time (or ad text adjustments, or device adjustments, etc.)! That's okay and natural, but it's why you see a ROAS (profitability or efficiency) dip when you surge spend, it's natural because we are locked into the limitations of our auction environment. The nature of rapid growth in PPC is that you have to be willing to spend money to grow your account, and then optimize back to ROAS.

Now with all that in mind, there is a crucial "ascension" aspect of this that needs to be brought out next.

(2) Your Business Should Grow in Top-Line Revenue Over Time.
Okay, so rapid revenue growth hits ROAS. Are we just stuck in this epic, plateaued good vs. evil seesaw battle the rest of our business lives? Absolutely not. This is where the "ascending" part of my definition of this metaphor comes in. While there is a seesaw shifting back and forth between revenue and ROAS, it should be simultaneously and steadily ascending over time.

Think of it more like a ride at the county fair. You remember those, right? We'd entrust our lives to rusty deathtraps hauled all over the continental United States with probably little to no safety inspections (can you tell I don't trust them?).

In this carnival ride, the "Ascending PPC Seesaw of Scaling", there is a hydraulic lift in the center of the seesaw that slowly lifts the entire seesaw up while it is tilting back and forth. So over time, while you are optimizing traffic and revenue to pull back on low performing targets and getting back to your target ROAS, you're also noticing that revenue is a little higher this month. And now it's a little higher the next month, and hey look at this, in month three we see our revenue grew and our ROAS is now stabilizing.

Wow, it worked... let's keep going!

In this admittedly overly simplified (I know... please don't tweet me to say this isn't exactly what happened in your account) model, I want to call out a couple of things.

The first observation being: it is common and expected for the recurring revenue drop to happen along the initial spend increase, but with a measured and controlled spend increase invested wisely, you can minimize the time it takes to get back up to your target ROAS... and it will likely be higher.

As long as that matches your goals, of course, you may not care about getting back to your 300% ROAS and rather want to keep pushing and hitting lower goals. That's fair, but just ensure that your expectations match your strategy.

What about the fact that in some months revenue actually dips? "WUT, UNACCEPTABLE, FIX THIS." Someone in the exec team (certainly not you) shouts.

Revenue should be going higher and not lower, always and foreeevvvvvvveeeerrrrrrrrr. The only scope of reality containing a constantly ascending entity is an exploration rocket shot out of the atmosphere never to return! Everything else in life has natural ebbs and flows, and this is no different. It's good to expect that, when you surge into new data points, you could actually see decreased REVENUE, as well as ROAS for a period, but here is where it is absolutely necessary to hire a solid PPC team/agency with a trustworthy strategy for growth.

If your keywords, audiences, bidding on their part, and the offer, landing page, sales process on your part are all locked down (remember what we discussed in a previous chapter about the limitations of PPC here!), then trust the marketing strategy and look to long-term growth. If you literally can't afford to spend more, then don't surge spend. Rather, maintain profitability to save money for a time when you are cash rich and in a better place to take a risk in surging traffic in a market. Sometimes the best thing (pandemic anyone?) you can do is pull back and maintain profitability while growing savings, rather than think you always need to be pushing.

It's worth bringing this out, because from the agency perspective, businesses who try to grow faster than they are actually able to manage with cash are the most egregious for panicking at revenue downturns since they literally can't afford any dip. As we've seen above, revenue dips can (and likely will) often occur in a hard growth stage. Business is hard.

The Power of Historical Data in PPC

If you're paying attention, you'll notice there is a crucial aspect here that needs to be revealed as we close: the power of historical data. An account with ten years of history is typically more valuable than an account just starting, and this is because you have ten entire years of data to use for bidding decisions, ad test decisions, new keyword theme ideas, etc.

You can't magically optimize a recently surged spend/traffic account to profitability because you don't have the data for the decision yet and need to spend more to get more data. It's not necessarily because your PPC agency is lazy, or ignorant (though, they could be either or both), but it's because you literally don't have the data yet. Data is immensely valuable, but the only way to acquire data is with time and/or money... and this generally stands in conflict with maintaining a hard profitability number. So I will reference again the point I brought up above which was: ensure that you have the cash to actually weather a rapid growth strategy. Business is hard.

I hope this has been helpful in considering how to rapidly grow a PPC account. At ZATO, what we're aiming for typically in an account (some accounts, admittedly are in unique places of growth or in a new vertical space and can grow profitability, rapidly as the exception) is a steady ascension over time that shifts between growth and profitability, all the while utilizing the growing data of power to grow top-line revenue over time. This, along with other marketing investments such as organic search, social strategies, and the all-important email marketing plan, can help establish a brand powerfully over time. This doesn't always align with every account or client's expectations, but then again, their expectations don't always align with reality. The trick is determining when something is unrealistic and when it's simply a big hairy audacious goal (to quote the excellent book, Built to Last[14]) that should be changed.

Business is hard.

---

[14] Jim Collins & Jerry Porras, *Built to Last: Successful Habits of Visionary Companies,* (New York: Harper Business, 1994).

12 ELEVEN REASONS YOUR GOOGLE ADS CVR MAY HAVE DROPPED

While conversations in Google Ads often revolve around bidding, targeting, and ad text or creative, the Landing Page plays a crucial role in Paid Search Marketing. Not just the landing page itself, but the elements of the landing page that lead people to take the action you want them to take once they have clicked on your ad.

How often do the people who actually click on your ads end up taking your desired action on the page? That rate between those visitors, and those who convert, is called the Conversion Rate. Even though Conversion Rate is a metric literally housed within Google Ads, it's important to be aware of the relationship to the landing page, since the reason people choose to purchase has more to do with what is on your website, your offer, or a variety of other business factors that may have nothing to do with your PPC ads at all.

There is an entire industry focused on CRO (Conversion Rate Optimization), so we won't delve deeply into those principles here. However, as it goes hand in hand with PPC, I wanted to provide you with some questions to ask yourself when you are analyzing your account and see a Conversion Rate drop.

If you've ever had a curious Conversion Rate drop on your website, you know that it can be difficult to identify the issue. Is this just a random seasonality thing or is there something more insidious at play here? I've found it can be helpful to have a checklist we ask our clients when investigating a Conversion Rate (CVR) drop, and I'd like to share that with you. Rule these things out and you're well on your way to figuring out what the actual issue is... or at least you'll successfully rule out what is not the cause!

Oh, and a brief caveat, I think it's really helpful to understand the difference conceptually between Click-Through-Rate (CTR) and Conversion Rate (CVR) drops. Simplistically (things are always more complex than this), CTR measures the efficacy with which a marketing or advertising message resonates with an audience when that involves a call to action to proceed to the website within the ad itself. If CTR drops, it's good to look at your ad or audience or placement itself.

CVR is a different animal because it reveals to you the percentage of people who visit your website, who then do, or do not, complete the checkout process and purchase from you. While it is possible for CVR to be impacted by marketing, we find that the problem most likely does not come from your advertising channel, unless something crucial changed in your targeting within that channel.

If you've had a recent drop in CVR, here are some questions with which to start:

1) Did a Sale Recently End or Begin?

Promotional events change user behavior, and you will likely see CVR rise during a promotional event (margin is a different story, so be careful about that). If a promotional event has just ended, this means you may simply be seeing a natural decrease in Conversion Rate on your site and in your campaigns.

2) Is This Just Normal Seasonal Behavior?

It's important to analyze previous year(s) performance to

determine whether you are simply entering a seasonally lower time of your business. Are fewer customers shopping for your products than before simply because of normal search behavior? Another way to investigate search interest changes is by utilizing Google Trends to monitor changes over time.

3) Was There a Recent Price Change?

I have a love/hate relationship with price changes in client accounts. I know, I know. They can create magical profit, which we all love! They can also harm conversion rate by causing fewer people to purchase. This rides on a number of factors, such as % of price change vs. consumer interest or desire in the product, and what price change most significantly harms CVR. I'll let smarter math people figure that out in your Finance team, but for the sake of analysis in this post: be aware that price changes can impact CVR changes, so have that in mind during your analysis.

4) Were There Any Changes to the Website of ANY Sort?

This is one of the questions we ask pretty quickly when digging into a change in consumer buying behavior on the website. Changes on a website can have two key effects:
- They can impact user behavior by changing the UI.
- They can accidentally disrupt conversion code or analytics code tracking.

How do you know which is the cause of your CVR? Well, this is why it's helpful to literally ask the open-ended question "were there any changes to the website, front or backend?" and then determine what could have happened with the reply. No change is too small, either. We've had situations where a client noted "well nothing changed... except this one product carousel was removed" or whatever it may be, and our suspicions lead us to believe that element that was moved or removed was actually influential.

5) Is Your Conversion Code Still Tracking Correctly?

Sometimes CVR changes in an advertising channel or analytics are not actual consumer buying changes, but they're literally just a conversion reporting issue. One of the things you can do is to first verify that your sales are still matching (or as closely as they'll match between two data sources) based on Analytics data and in your actual store tracked data. If your store data and analytics data are both reflecting a drop in sales, then something bigger than tracking is likely the culprit. If your store data shows sales are as strong as ever, but your analytics platform has seen an overnight drop... then a conversion tracking issue is almost definitely the culprit (see Question 4!).

6) Is This Change Replicated Across All Channels, or Only One?

This is another one we look at pretty quickly. You can identify channel behavior quickly in your Analytics platform, and what I like to do is look at a very simplistic pre/post event time frame. Let's say the CVR drop happened six days ago, so set your timeframe to the past six days and compare that with a previous period of six days. What you are looking for is a strong variation in CVR change between channels. Is the CVR drop obviously limited to a single channel such as Google Ads or Facebook? Those are good indications that the specific conversion tag for those channels may have been dropped somehow from your checkout pages. Or, at the very least you can dig in to determine if targeting in your ad campaigns caused a shift... or something. But it's very helpful in determining whether this is a bigger issue or just something limited to a specific channel.

7) Did Other "Hidden" Costs Such as Shipping Rates Change?

This is a tricky one because you really have to start digging; however, a customer often evaluates all costs as part of their purchase decision (especially for more impulse-buy products). If your carrier rates have recently increased for shipping, it could be discouraging people from purchasing. After all, why would you want to buy a $12 hand lotion that costs $12 to ship when a similar "looking" hand lotion on Amazon is $14 with free shipping? Don't @ me cause your hand lotion is so much better quality, because I'm talking about the mind of the consumer and your CVR drop here!

8) Does The Site Speed Load Slowly, Or Has the Speed Changed Recently?

Site speed is an increasingly important factor, again, especially with more impulse-buy products. Consumers may have seen your Instagram ad, thought "what the heck I'll give this a try" while they're looking through reels for hour #2 in bed... but if they can't access your site, they might not purchase. Do consumers have that short of an attention span?? Are they that fickle? Yes. Yes they are. According to an article published on[15], just shortening your load time by 1 second can result in a 27% increase in mobile conversions.

Yowza.

9) Have Competitor Prices Changed Recently?

This is another one you have to dig for, but it can impact your conversion rates in a big way. Competitors lowering prices can cause unfortunate CVR dips because we have found this behavior often doesn't actually reduce traffic and clicks from ads, but it does lower CVR. This has the unfortunate result of costing you money to continue to send curious people to the site, who then bounce and head to your competition. Ouch! You're in charge of your own pricing, of course, but it's helpful to at least know if this is the cause so you can have a difficult margin conversation if that is the case. One more thought on this one is that a retail er breaking MAP on a specific product can really damage your CVR since they

---

[15] Ainsley Wilson, "Why Site Speed Is So Important For CRO, SEO, and Loyalty," blog post, 2022, *Shopify Blog,* accessed 2 February, 2024, https://www.shopify.com/blog/site-speed-importance.

might be trying to sneak in a lot of sales quickly before they get the cease and desist from the manufacturer, so be aware of that (and your MAP agreements!).

10) Is there a Difference in Conversion Rates Between New and Returning?

I like to investigate this in terms of user location in the purchase funnel, though I don't think I'd say this is a real common finding. Still, it can be one to dig into. Perhaps there is a bigger issue at play here in how you are positioning your products in advertising or on your website to different audiences. Perhaps you have positioned yourself well with new customers, but are struggling to convince previous customers to purchase again. It is worth determining this so you can test new UVPs and even new assets and creative.

11) Is A Key Product(s) Out of Stock?

Make sure to ensure all of your core products are actually still in stock, or that there are no issues with them. Checking into which SKUs have actually decreased can even help lead you to other insights with more digging, such as whether competitor pricing changed in the same auctions as your core products.

One of the reports I like to run when auditing any sort of performance dip is to analyze product sales in Google Ads (this would be for Shopping or Performance Max campaigns only) to determine if there was a drop-off on any specific SKUs. Simply start your report timeframe from the day of the drop through today, and compare to the previous period. Then make sure to sort by highest selling products in the previous period. It's not ultra scientific, but you'll be surprised how specific products can sometimes jump out at you.

All that to say, you can't always stop people from purchasing, but with some insightful analysis, you may be able to identify the source of the bleeding and triage the wound before it gets too bad by asking these 11 questions.

13 A TALE OF ATTRIBUTION WOE AND COW CRAP

Ahem. A story to illustrate digital attribution. This is a story I wrote years ago, but overall, I think it holds up. Enjoy!

Janet likes cows.

Ever since Janet first visited her grandparents' dairy farm in small-town Wisconsin, Janet has had pleasant thoughts of her bovine friends. As she got older, Janet could not visit the farm as much as she would have liked, so she compensated by purchasing cow memorabilia off of the internet.

If you could put a cow on it, Janet would buy it. One day, Janet's friend told her about a new site, Cow Crap, LLC (http://cow-crap-it-all.com). Apparently, Cow Crap would take anything you mailed to them, stamp a cow on it, and send it back.

She was delighted and immediately Bing-ed the term (Janet used Bing since it was installed on her work computer by default and she didn't care enough to change browsers). She found an ad by Cow Crap and clicked on it. Seven hours later, she left the site and headed back to work. She couldn't get her mind off of Cow Crap, however, and she pondered which of her items to send in to be cow-stamped.

The next day, Janet woke up and checked Facebook immediately... upon which she saw an ad for Cow Crap! She couldn't believe her luck and wandered back into the site.

Over the course of the next five days, she would visit Cow Crap 12 more times. 4x through Facebook Ads, 3x by typing the URL directly into her browser, 2x through the emails she was beginning to receive, and 3x through ads in Bing.

After much deliberation, she ultimately decided to send in her bed frame to be cow-stamped; so she typed in http://cow-crap-it-all.com, clicked the CTA, and finally gave Cow Crap her money.

Now, "attribution" answers the question, which channel gets the credit for this sale?

Here is where things get complicated in many Digital Marketing reports because Google Universal Analytics reported sales with Last Click Non-Direct attribution. GA4 now utilizes a machine learning cross-channel data-driven attribution model.

When the Cow Crap Data Analysts would look at their Google Universal Analytics reports in the old days, they would see that this sale came by way of Google Ads (at least, the last click was Ads).

In this older model of last click, they had noticed a lot of Google Ads sales coming lately, which makes them feel warm and fuzzy since they are spending so much money on Google Ads. Their CMO, on the other hand, is fresh out of her previous people management job (Cow Crap felt her "strengths" for this job were her people skills as opposed to her marketing skills), and she has been looking at Channel Reports. "COW CRAPPERS!" She calls, demanding a meeting. "I have noticed that we spend far far far too much money on Facebook Advertising." When I look at where our sales are coming from, it is clear they are not coming from Facebook Advertising." "But…" began a low-level analyst. "No buts," She interrupted. "We need to put the money where it gives the best return.

I want you to pull our Facebook Ads and send that budget over to doing all we can to increase our referrals and Google Ads traffic; that is clearly where we make the most money. I want creative ideas on radio advertising, city bus ads, billboards, whatever you need to do to get the word out in a way that will increase referrals. Oh, also, boost our Google Ads budget because that sends sales."

Cow Crap did this, and unfortunately their sales began to dry up.

You see, while Google Ads branded keywords sent many sales like this, and thus Google Ads got 100% of the credit for those sales, it was by no means the entry point for most users (think about it, they typed in the brand so clearly they had to be familiar with it already). When they pulled their Facebook Ads budget, they ended up strangling (quite gruesomely, I might add) the Top of their Sales Funnel and drying up their sales.

They went out of business, and everyone cried and cried. It was sad. The moral of this story?

People often do not make a buying decision with one visit, so make sure you understand the journey of your buyers before you make a decision you will regret.

Nonsensical stories aside, understanding attribution (note, I didn't say "figuring it out") is an essential part of any digital marketing strategy. Unfortunately, it is also an evolving industry… which means there is still a lot of guesswork and change involved.

As we have already mentioned, GA4 now utilizes data-driven-attribution (DDA) in which a machine learning algorithm determines how best to dole out percentages of a sale to different channels. Even this carries difficulty, however, since it is rigidly assigning value based solely on knowable data.

And with this in mind, I'd invite you to read the next chapter.

14 AVOID THESE TWO DITCHES OF DIGITAL ATTRIBUTION

I spent my early driving years navigating Wisconsin and Minnesota winters. Here's the deal, when you are driving in the middle of a white-out snowstorm on a back country road, it doesn't matter how slowly you are going as you can't tell what is road and what is ditch or nearby field. During those times, you aim for the middle of the small two lane road as you seek to avoid either the ditch on the right or the left. If you see an oncoming car, you both slow way down and aim to pass each other on the correct side, while also navigating the ditches.

On one such occasion as I drove from Minnesota to Montana to visit my (soon-to-be) in-laws, I remember getting out of the car with hands almost frozen in a clenched position after gripping the steering wheel so tightly for so long!

When it comes to attribution, I believe there are two ditches that need to be avoided by the marketer as you aim to just keep the car of analysis on the road. The first ditch is the more obvious one: it is the ditch of attribution ignorance.

Attribution Ditch #1: Attribution Ignorance

This ditch is the ignorant belief (whether accidental or

purposeful) that a user journey is not a complex sum of varying touchpoints with the brand. It often reveals itself as an obsession with last click tracked sales ROAS (even if that is not the attribution model used!). It's less of a rigid loyalty to an actual last click attribution model and more of a last-click attribution mindset or culture. "Did this channel get us more sales, and if so, let's give it more budget." In our complicated digital sphere,it is crucial to be aware of the difficulties of attribution when setting budgets and assigning ROI properly, and it is no longer an excuse to ignore attribution.

Admittedly, this ditch has increasingly been called out and warned against successfully in the PPC industry. There is still a long way to go, but attribution awareness has continued to grow over the years. I find that even business owners and CEOs are hungry to unpack the puzzle of the attribution enigma in their accounts, no longer just the marketing department that is aware of the concept of attribution models.

Unfortunately, to avoid the ditch of ignorance, some may veer wildly to the other side of the road... and head directly into the ditch of arrogance.

Attribution Ditch #2: Attribution Arrogance

Whereas attribution ignorance is undervaluing the knowledge that attribution can bring to your business, attribution arrogance is overvaluing the knowledge that can be gained.

It looks at a specific model included in your analytics software of choice, assigns X% of value to each source, and confidently sends a report to the client, "thus hath the mines of mystery been plumbed, and henceforth shalt the budget be setteth." Nobody is more guilty of this than Platforms themselves. In Google, Data-Driven-Attribution is often trumpeted as the solution to all your attribution problems rather than simply one way of viewing your performance attribution.

This is a ditch because it communicates to the client that attribution is simplistic and easily solvable based on the specifically tracked data to which the model has access, requiring only some specific, magic formula in order for infallible ROI measuring awesomeness to be grasped. By the way, it's ironic that the

Platform reporting on their channel has "clearly figured out" that "perfect formula", and it coincidentally gives their channel quite a bit of credit.

A good marketer uses data. A great marketer uses data to take action on what she believes to be true that has not yet been proven (and sometimes can never be proven), regardless of the attribution model's simplistic assignment of value.

A specific attribution model can only take us so far in determining true success of a channel, and therein lies the inherent weakness of attribution.

The glaring weakness of attribution is none other than our inability to accurately track human emotions. By that I mean, attribution will always be limited to the data it collects. It simply calculates a percentage of credit based upon aggregated data of users who made certain steps in their journey and then making assumptions of value based on that specifically tracked click behavior.

Yet crucially, an attribution model cannot identify when the actual decision made in a purchase happened in the mind of the customer.

Allow me to illustrate this with one of my favorite characters (he was my favorite long before he was made popular by Benedict Cumberbatch!), Sherlock Holmes. Sherlock Holmes is a master of deducing facts in order to solve a case, but not every fact and not every deduction holds equal value in the resolution of a case. For instance, he may discover fibers on the floor that lead him down a mental path, and then he might interview a witness who lies about a key piece of evidence, and then this might cause him to visit the moor itself, whereby he will put the finishing touches on the case (it was the butler, with the candlestick, in the smoking lounge!).

Yes, attribution can answer the question: "which factual interactions led Sherlock Holmes to solve the case?" But attribution can never properly weight those. I do realize never is a strong word, but I stand by it.

For instance, the analytical compiling of event facts and user behavior data cannot reveal the fact that it was the witness lying that gave Holmes the *most internal suspicion*, which led him to pursue the case more intentionally and thus visit the moor, leading to the resolution. Without Holmes (or, Watson for that matter, or really Sir Arthur Conan Doyle) actually telling us what went on in his head, we cannot know his intentions and how they were impacted by each interaction.

The weakness in using this as an example is that we, as readers, can see into the head of Holmes, so we are brought into the decision in a way that does not happen with online user attribution.

While you can track user behavior on your site, and you can identify and fire various events to identify who did what, when on your site, you still can never actually know which channel caused the most "credit" for a sale in the mind and emotions of the user. This is absolutely crucial because we analyze attribution data in a percentage model. A Linear model doles out equal percentages of credit to the channels in the entire user journey, and a Last Click model doles out 100% of the credit to the last channel to send traffic, and so on.

However, these are giving out credit as percentages based solely on timing of sessions and not on how we as humans actually make decisions... with emotion, with logic, with reason, with desire. At the very least, one change that needs to happen immediately is fewer boldfaced ROI claims from attribution and more honest communication with clients into the actual state of things.

Be less concerned with finding a 100% perfect attribution model and more concerned with diving deep into a partnership built on trust that will allow you and the client to adapt over time as you continue to experiment and tweak their attribution model based upon source interactions over time and aiming at more directional focused decisions for truly incremental sales. *(My editor told me that was a long sentence, so instead of fixing it I'm going to suggest you go read it again, slowly this time... pondering it.)*

So to close, as we think about attribution, I'd like to warn us not to run from one ditch into the other. Attribution is evolving in digital marketing, but our understanding of it needs to evolve as well. We need to stop simply asking "how were the sources arranged in this transaction?" (that's a great place to start) and instead begin asking immediately after the first question, "what can I learn about my customer's emotional orientation towards my brand in each channel?"

Finally, we need to be okay with not having attribution 100% figured out. We can't know it perfectly. We can never know it perfectly. Take a deep breath, repeat that to yourself, and then do the best you can in your client and maybe even, *gasp* rely on your gut sometimes.

15 THE DIGITAL ATTRIBUTION BUBBLE

On November 6, 2019 (immediately before the Ecommerce holiday season), Jesse Frederik and Maurits Martijn took aim at the $273B industry of Digital Advertising and wrote an article in The Correspondent entitled: "The new dot com bubble is here: it's called online advertising."[16]

I remember first seeing the article and glancing over it but not actually reading it for a week or so, since I was traveling a bit in 2019 for conferences and trying to stay up on client work and life in general. I finally had the chance to read the article in depth in the Minneapolis airport on a layover (when you live in Billings, MT, you only have a direct flight to about five cities), and the skillful writing, and significant assertions captured my attention immediately.

I found myself nodding my head vigorously at much of Frederik and Martijn's article, specifically as it called out the digital advertising community's obsession with chasing tracked profit at the expense of incremental value and actual new customer growth. I'm in hearty agreement with that, and I even found myself hoping this would help push our industry to higher heights in certain areas.

---

[16] Jesse Frederik and Maurits Martijn, "The new dot com bubble is here: it's called online advertising," blog post, 2019, *The Correspondent,* accessed 2 February, 2024, https://thecorrespondent.com/100/the-new-dot-com-bubble-is-here-its-called-online-advertising.

That being said, I heartily disagreed with the article's premise that online advertising is itself the issue. The authors blamed the failure of marketers and engineers to actually demonstrate incremental value on digital advertising itself, rather than on the tactics devised from an improper understanding of attribution. The real problem in digital marketing right now isn't that digital marketing exists, but it is our belief that we can track everything.

Here is what I mean: *digital advertising is just advertising.*

It's not the greatest thing to ever happen to marketing, and it's not a bubble. It's just advertising. It's doing what marketers have done for years: utilize a specific medium to grow a brand (and thus sales) by getting the right message in front of the right people at the right time.

Discovering, as the authors did, that people at eBay running most of their budget into their own (exceptionally powerful) brand terms are surprised to learn they don't see incremental value isn't the nail in the coffin for advertising that the article suggests. eBay was just running PPC poorly.

Ironically, PPCers in my sphere have long written on the poor eBay PPC program management evident even from the public eye ("used babies" in titles thanks to DKI, anyone?). This is why it's crucial to point out the thing I believe the article alludes to but doesn't actually identify as the actual bubble: that is an improper understanding of attribution, and how that establishes misguided tactics for paid search accounts that fail to build brands and add incremental value.

In other words, I am positing that paid search advertising itself has not failed; it's that an understanding of how to use paid search advertising as part of an integrated marketing mix for individual companies has failed. Improper use of attribution has led to an obsession with directly tracked results that over time do not build a brand and incremental sales. They simply retarget (not necessarily remarketing, by the way) the same users already in the sales cycle – ad nauseam.

In this regard, I would suggest that conversion tracking is as much of a curse as it is a blessing. Conversions, of course, are the specific (tracked) action a user takes to accomplish the goal of the advertiser. When you can track what source led to a sale, you begin to think you have an understanding of how your consumers purchase, and you begin to invest more money into that source. But what if that source is only one piece of the puzzle... especially if it's closer to the bottom of the buying funnel, meaning much contact has already been likely made by your company?

When you think you can track everything, you begin to shift your time, resources, tools, and reporting to making your trackable KPIs grow, rather than building and implementing the tactics to accomplish an actual marketing strategy within your digital channel.

If your paid search strategy focuses solely on sweeping up those bottom of funnel clicks and sales (which is what you're tempted to do with a last-click attribution model in a demand capture ad channel, which gives 100% of the sale credit to the last source to send you the sale), then yup... you maybe won't see much damage (at least initially) in pausing paid search.

To be clear, certainly, there are times in competitive industries (especially with startups who don't have more advanced marketing channels built out yet) where using paid ads to initially catch those bottom funnel users is a sound tactic, but this should not be the primary way advertising performance is measured.

I'm going to complicate things even further. You may be reading this and vehemently agreeing with my concerns. "Death to last-click attribution!" you cry. C'mon, it's 2024. We all know that.

So, let me push into this even further. There is no perfect attribution model, and with privacy awareness increasing and dark traffic in a continually strong space, this means you can't really trust a more complex attribution model either.

Why did a person visit on their third visit and decide they "loved this brand and had to have one" because of "just the right" emotional experience? But, then they didn't actually purchase until their seventh visit, 12 days later? Who knows. That's not something you can track, and we will be better marketers once we embrace these limits of attribution.

Attribution will always have limitations, which is why in some ways attribution (improper or proper usage) itself may be the true dot com bubble the authors are trying to sniff out. I'll reiterate: advertising isn't a bubble, but it's a tried and true business service. But the thought that we can identify direct sources of performance and thus shift the entire budget to those sources based on suspicious data? Yeah, I'd agree that's a bubble, and that's attribution.

Regardless, as long as we keep chasing solely after tracked individual channel success and building digital marketing strategies (selecting keywords, ad text choices, locations, devices, audiences, demographics, etc.) without thinking beyond individual channel success, then we will continue to struggle to build brands and see incremental growth.

Only when we as paid search marketers strategize with the other channels to build a marketing strategy targeting the right message to people at the right place in the funnel (selecting unique keywords, and channels and campaign types for those, of course) will we begin to get beyond tracked ROAS as our primary KPI and focus more on overall brand growth across all marketing channels.

If what you're hearing scares you because it sounds risky, well then, I think you're picking up what I'm dropping. Real, authentic marketing that builds lasting brands has always been difficult, risky, time-consuming and expensive (with a healthy dose of luck). I don't think digital advertising is the dot com bubble. I think our belief that we can track everything is the dot com bubble.

And what's more, with the incoming onslaught of privacy regulations removing devices throughout a funnel, it's getting more and more difficult to track users all the way through the funnel now anyway. In other words, whether you agree with what I'm saying or not, the reality is that we're all going to be forced into more of a traditional marketing mindset soon anyway as conversions become more modeled and attribution models take more and more liberties in assumption.

Time to get back to marketing to worry less about tracked profit and to build a brand.

16 WHEN DATA & AUTOMATION DISAGREE

"The year of automation is here!"
"Don't let your team manually do what should be automated."
"The one who still manages bids manually in Google Ads is wasting her time."

These and other like-minded sentiments are being uttered by PPCers who just a couple of short years ago were (rightfully) hesitant about incorporating too much automation into their systems.

What happened? Call it an evolution of the system (machine learning, by definition, gets smarter as time passes), evolution of our thinking as Paid Search managers, or most likely a mixture of both... but the world of PPC has certainly merged into the lane of no return on the road to automation.

In my paid search agency, ZATO, we do a bit of work with automated systems in Google Ads, both in moderated, human-controlled fashion as well as in near fully-automated systems such as Google's Performance Max (PMax) Campaigns released in 2021.

I have engaged in debates on PMax data loss and automation and have written many articles on the topic of PMax (you can view those articles I've written on my website, if you so choose). I am especially interested in PMax automation because I believe it is the culmination of "the new Google Ads", along with the more demand creation focused "Demand Gen Campaigns" from Google (PMax itself tends to be more Demand Capture by aiming at lower funnel traffic).

That is, Google wants to utilize more automated systems such as audiences or product feeds to allow them to control more of the placement and bidding of ads than the "old school" method of keyword targeting. In this regard, I think there are crucial aspects to automation that the PPC industry (and in many ways, the broader digital marketing industry) needs to sit up and take notice… before we are too far down the path of obscure data.

Based on this, I want to dig into four crucial and somewhat contradictory elements of automation of which our industry needs to be aware, talk about, research, and determine the best route forward for us PPCers. This is more a chapter intended to get us to think as an industry, with unanswered questions and ponderings.

Why Is This a Crucial Conversation Right Now?

It is a crucial conversation to have now because the platforms want to "obfuscate" (literally, this was the word chosen by Google GM Sissie Hsiao[17]) the data because it is in their favor to run everything.

The argument is that obfuscated data will make for better machines. But shouldn't we pump the brakes a bit and consider this? I am certainly no Luddite, but the speed with which we are launching into fully automated systems in PPC suggests we need to slow down to contemplate a few things. Of course, we can't go too slowly so as to lose out on innovation, and I think the time for contemplation in this regard is now or never. If we don't discuss this and help our industry develop strong, guiding convictions on the various aspects and nature of data and automation, then we are giving the Platforms free rein to do as they please.

In this regard, the advertising platforms (primarily Google, Microsoft, and Meta, in my mind) have already communicated through public statements and private conversations I have had

---

[17] Sarah Sluis, "Google GM Sissie Hsiao Is Planning For the Next 'Jump Forward'," blog post, 2020, *AdExchanger,* accessed 2 February, 2024, https://www.adexchanger.com/platforms/google-gm-sissie-hsiao-is-planning-for-the-next-jump-forward/.

with different Google representatives at various conferences that their desire is to: Control every aspect of the automated process. Keep that data hidden to prevent outside influence on the algorithm.

I think this is problematic for a variety of reasons, as I have written elsewhere in my concerns on the now deprecated Smart Shopping Campaigns.[18] My hope is that conversations like this will help automation continue to develop and do great things while maintaining adequate human oversight and process transparency.

The purpose of this chapter is not to reconcile differences (I don't have enough background in coding to do that well), but to reveal difficulties (even contradictions), to encourage conversation, and to get the right people thinking who can work actual change in this regard. The conversation is the win for me here and why I thought it was an important chapter to include in this book.

Therefore, if you disagree with something I say next, great! Talk about it; write about it. Rather than allow the platforms (with the most to gain) to completely shape the automation process and conversation, let's discuss the following elements of automation (and more!) and help move our industry forward.

Two Sets of Necessary, but Contradictory Elements of Automation

<u>Kirk's First Opinion: The Advertiser Has Certain Rights to Access the Data</u>

The first necessary element of automation I believe is crucial here is that the one paying for the data has some right to access that data. How much right? I don't really know.

In its third party policy with advertisers, Google itself claims, that there is certain data necessary to display to the one actually paying money to Google for Ads Program usage. In other words, a

---

[18] Kirk Williams, "Should Smart Shopping Include More Data? My Rebuttal to The Rebuttal - Part 6: The ZATO Guide to Google Smart Shopping Campaigns," blog post, 2021, *ZATO Blog*, accessed 2 February, 2024, https://zatomarketing.com/blog/should-smart-shopping-include-more-data-my-rebuttal-to-the-rebuttal-part-6-the-zato-guide-to-google-smart-shopping-campaigns.

Paid Search agency like my own must share certain data with advertisers in order to align itself with Google's Third-Party policies.[19] Of course, Google is masterful at keeping clearly defined language from sneaking into this policy. It's not like they claim "search terms data is significant" for agencies. However, what I gather from this is Google's admittance that an advertiser has certain access to data rights because of their payment to the Platform to use the advertising service. In other words, the question isn't "do advertisers paying Google have certain access rights to data"? Because as I have shown above, Google themselves believes there to be some level of access rights.

The question is "to which data do advertisers (paying money to Google) have access rights"? It may be that some businesses are impacted more significantly from losing 30% of search terms data than other businesses. That's the nature of business, and in my opinion, it is precisely why the best solution is for Platforms to cease the total obfuscation of data for the sake of more automated campaign types that may or may not work as well for all advertisers.

The Problem With Kirk's First Opinion: Not All Data Is Actionable or Helpful

Here's the challenge with the above, and thus why it is seemingly contradictory and messy... What data do you have a right to as the one paying for it?

All of it everywhere?

Or only what is necessary?

And in that case, who determines what is necessary?

You want all of the data, everywhere? Then go buy a server farm(s) for the mounds of junk data that you can't do anything with... for the loads of this, that, and the other data points you'll never, ever, ever use.

Unlike the trumpeted core value of "big data", the one with the most data doesn't actually win anything. It's the one who can use data correctly who will win. We saw above that Google calls this

---

[19] Google, "Google third-party policy," Google, accessed 2 February 2024, https://support.google.com/adspolicy/answer/6086450?hl=en.

necessary data the "right information" for making informed decisions... but that's complex as well.

One advertiser thought Average Position was crucial for making informed decisions, while another thought it wasn't necessary and doesn't really miss it. Who is right? Advertiser A may very well have found value and utilized Average Position well for their bidding, and Advertiser B may very well have ignored it for Search Impression Share and the like. They may have both very successfully made their accounts money based off of these conflicting views of "necessary data."

So, who is right? Well, in a way, both of them. They each have their own process for utilizing a certain datapoint in building a successful account. The PPCer will say "it depends" to describe situations such as this. But according to Google, in making decisions such as blanketly removing Average Position from access for all advertisers, whatever Google determines is important is the necessary data that all advertisers get in their accounts. Do you see my concern here? In not discussing this further, we are handing the platforms the ability to not simply automate our campaigns but to determine what data is actually "necessary" for us to make "informed decisions". "Just take your label-less medicine and trust us... it will help you. We're pretty sure. Until it doesn't."

A combination of these two realities makes sense to me: Increased advances in automation for those who don't care about specific control and data access, plus full access to the data-sets for those advertisers who believe they need it. This will likely take some work and is certainly not the most efficient way to do business for Google, but it is arguably the best way to do business. At some point, doing business with so many unique entities isn't solely about efficiency that utilizes averages. Automation shines with grouped efficiency and averages...not shared and thus manually managed data.

That is also why I believe social audiences are becoming more popular over and above search terms data with Google (grouped averages are easier to automate well), even though the search term and keyword is why Google has always shone brighter than other marketing channels and is what continues to draw new advertisers to the platform (see previous chapters on this).

The problem for Google is the businesses that don't fit into those averages but also have a right to certain data since they are also paying for the ad program, especially small businesses. My question is: isn't that the cost of doing business with people, especially lots of people? The small realtor in Billings, MT, paying Google pennies is part of the advertising ecosystem and has equal data access rights as the billion dollar travel entity. An engineer's brilliant plan for an automated system that would work for the billion dollar travel entity, but result in the failed account of the small realtor, isn't actually the right way...even if it's more efficient for some automated program.

Doing business with so many different people is messy and can't always be boxed up neatly into a closed-system, automated process. Sometimes the best thing you can do as a Platform is to understand that profit and efficiency are at risk and to seek the solution that allows for the most data points to be accessible so all paying customers can utilize your advertising solution. At least, that is my opinion. At some point in the very near future, "data access rights" will likely be decided in a high court somewhere, and I guess we'll all watch with fascination and trepidation.

<u>Kirk's Second Opinion: The Algorithm Works Best When It's Given Proper Guidelines</u>

Well, that got complicated. What about the next set of contradictory but factual opinions on automation?

Let's start with the opinion that resonates with us advertisers: machine learning needs humans for ongoing feedback.

Now admittedly, I've heard this concept spoken of positively by platform reps as well. Many people agree that some sort of human guideline is necessary for most (all?) forms of PPC automation. Machine learning isn't artificial intelligence (no matter how many people put AI on their websites). It relies on connected paths from past data points to make the best decisions. This also means it needs guidelines to *keep* it pointing in the right direction. Without guidelines such as accurate conversion tracking and efficiency targets, the machine will run after the "wrong" targets. Wrong here does not necessarily mean it is misguided based on the information it has, but that it is wrong for your business.

"Make us money by spending money" told to a machine will release it on endless audiences looking for the chance to do just that. It might work, but it could burn through a LOT of money and time while doing so.

"Make us a 400% Return on Ad Spend, while spending $100 per day" as an added guideline suddenly sets essential "out of bounds" around the machine's intended path. There are definitions for what success looks like, and by measuring past success from certain audiences, the machine can more easily identify potential wins based on that history plus guidelines.

The Problem With Kirk's Second Opinion: The Algorithm Works Best When It's Not Being Changed Unnecessarily
What could possibly be the problem with the above statement, especially since many agree that some sort of guideline is essential for automation?
The problem is back to the definition of what is actually necessary. The worst thing for an algorithm is to get bad data. The second worst thing is to change it when it doesn't need to be changed.

Manual tweaks have killed many an automated process.

A great presentation on the concept around ad testing in paid search was given by Martin Roettgerding at HeroConf London a few years back.[20] In it, he demonstrated how many human ad testers often make decisions too quickly with not enough data, whereas an automated process is more likely to see results average (and rise) over a longer period of time. In other words, we humans can make changes unnecessarily, or without enough data, and mess up the algorithm. Platform engineers are rightfully concerned about giving an element of control over to humans who don't fully understand the process. It's a difficult friction in which we can both agree that humans are necessary and yet also agree that humans can mess it all up. Even intelligent, professional, well-intentioned PPCer humans.

---

[20] Martin Roettgerding, "Debunking Ad Testing Part 1: Statistical Significance," blog post, 2018, *PPC Epiphany Blog,* accessed 2 February, 2024, https://www.ppc-epiphany.com/2018/10/23/debunking-ad-testing-part-1-statistical-significance/.

So what do we do about all of this?

Here is where I would like to again make my request to the advertiser community. We need to see the danger with ad platforms solely defining the "terms" of what data is necessary and beneficial, as well as obfuscating the process of the algorithms.

Going back to PMax Campaigns (since it is often where my head is at these days), I have heard the previous concern that humans will mess up the algorithm by making changes used by Google in order to prevent exceptionally helpful data like which search terms cost the most without driving conversions or Placement reporting metrics.

If we allow these decisions to continue unchallenged, the platforms will ensure "necessary information" and "essential guidelines to the automation system" are defined solely by how they want to define them. The platforms are in the business of making money ("shaking the couch cushions" anyone?[21]), which will inevitably impact how those definitions are ultimately established, and we would be foolish to expect anything else.

When the platforms stand to benefit from any party in the auction and they are the only ones who have any idea of what is actually happening in the auctions, we shouldn't be naive enough to expect them to do what is best for everyone else. If anything, perhaps it would be better for automation if as much data as possible remained available and advertisers thus maintained a greater element of control… even if it slows down the machine learning growth curve and introduces the potential for disaster at the hands of the wrong human guides.

Why is that… for the sake of avoiding the inevitability of our machine-overlord futures? No, I think open data within the algorithm process and more control over of that process will do a few things:

Allow for different strategies and tactics by different advertisers for the data, rather than forcing everyone into the same box

---

[21] Marketing O'Clock, "Shakin' The Cushions: Google Exec Admits to Bid Increases To Hit Quarterly Goals," recorded podcast episode, 2023, *Marketing O'Clock Blog,* accessed 2 February, 2024, https://marketingoclock.com/episodes/shakin-the-cushions-google-exec-admits-to-bid-increases-to-hit-quarterly-goals/.

(remember the Average Position example shared earlier). I think there are broader business benefits to be had by data that Google doesn't realize, which can help a business grow in its totality... rather than just limiting the data to be used (and thus hidden) by Google Ads alone. And a happy, growing brand is more likely to continue to invest in a channel if there are other benefits (such as data) it receives from that channel, rather than JUST trackable performance alone.

Allow for responsibility to be placed on the advertisers. Perhaps Google has taken too much responsibility that is not its to control in individual ad accounts by forcing averaged, automated control onto all advertiser accounts?

Avoid the inevitable unethical behavior that will result from entirely closed systems where literally billions of dollars are at stake. When that much money is at stake, and in an entirely closed system with only insight into it from internal teams (for example, consider how easy it would be for Google to game a system with a data-driven attribution model applied to PMax campaigns when they control everything including how sales are attributed), it's impossible to imagine a future where someone at some point doesn't use it for evil. It's just too big of a temptation for humanity (more on this in the appendix!).

Transparency and process can slow progress and efficiency down at times, but it can protect that progress and efficiency at the same time. So, by all means, find problems with this chapter, discuss it and point out its logical holes, but for goodness' sake, let's shine increased light on the decreasing lack of data being handed to those paying for it in paid search at this time.

17 HOW TO TEST WITH LOW BUDGETS

Let's face it, with rising CPC costs (thanks in no small part to Google's auction tinkering to hit their revenue goals; see updates on the current US antitrust hearing against Google at the time of writing this chapter[22]) Google Ads clicks are increasingly expensive. Sure, published CPC trends by Google often show a decrease, but that tends to be because additional lower cost (and thus, lower value) channels are included in aggregate sums. For the brand keeping an eye on actual Search or Shopping CPCs in specific searches over the years, things tend to be more expensive than they were just a few years ago. This means, among other things, that it's more difficult for the small advertiser to win at Google Ads.

Difficult, but not impossible.

Some may argue that SMB PPC advertising has always been difficult to do well on Google Ads, and they would be right. It's why, in part, we created our MAKROZ SMB PPC Advertising brand! We saw the need for a good solution to advertise lower budget accounts on Google Ads that wouldn't break the bank.

For the good marketer, advertising on Google Ads with a small

---

[22] Nicola Agius, "Google quietly increases ad prices to meet targets, claims exec," blog post, 2023, *Search Engine Land,* accessed 2 February, 2024, https://searchengineland.com/google-quietly-increases-ad-prices-targets-432155.

budget is possible, but it requires some specific methodology and work, especially as it relates to testing new opportunities in the account. Based on things we've learned in our MAKROZ SMB PPC Program, I wanted to share those findings with you.

Demand Capture, Not Creation for SMBs

Remember, Google Ads is demand capture, which means the demand must be generated elsewhere. This is why at some point in your SMB account you will likely hit a plateau and struggle to grow beyond that. You find yourself showing up in the same number of searches for the same keywords you're targeting, at the same traffic and Click-Through-Rate (CTR), and at the same Conversion Rate (CVR). You may be able to increase conversions somewhat by optimizing your landing page or rethinking a different offer (really important things to do, by the way!), but overall you will not see increased traffic on core terms unless some external marketing effort (by you or someone else) causes additional demand for those keywords. This is the unfortunate, but important reality of advertising in paid search.

It Takes Smart, Purposeful Testing

The other primary way you can purposefully grow your lower budget SMB Google Ads Account is by intelligent testing to find (and capture!) new pre-existing opportunities. The problem with this, of course, is your darn limited budget! Typically, in a larger, more established PPC account, it's wise to spend around 80% of your budget on tried and true keywords and targeting methods in your account and then to spend around 20% on testing. This is more difficult in an account where you're spending $2,000/mo or even $500/mo. You don't want to spend 20% of your small budget on testing, because, (1) that 20% loss in your guaranteed targets will hurt big time, and (2) 20% of a $500 budget is $100. If your clicks cost you $1.50 in avg CPC, then your testing is only analyzing around 66 additional clicks PER MONTH.

Only 66 clicks per month!

It's nearly impossible to test that successfully. It could take months or even years to get enough data to feel confident in making a wise decision.

So, you're stuck, right?

Not necessarily! Here's where it may help to adjust your testing strategy, and I'd like to show you how we think about that in my agency.

The key is to:
1) Identify actually significant, successful targets in the account and free up existing budget by eliminating everything else
2) Invest in a smart burst-spend testing strategy

Let's start by looking at number one above.

<u>1) Identify actually significant, successful targets in the account and free up existing budget by eliminating everything else</u>
When it comes to Small Budget Accounts, we have to work aggressively to reign in our desire to target all the things. It's absolutely essential to (a) focus on what 100% absolutely unequivocally will work every time and (b) test everything else in purposeful bursts.

In other words, if you are targeting 100 [exact match] search keywords in your account, but only 50 have ever spent anything at all and only 15 keywords have driven enough traffic and sales to guarantee their worth, then you need to focus SOLELY on those 15 keywords and begin testing everything else in bursts. This will help you find additional money you didn't know you had to allocate for testing.

This can be the same for audiences, campaigns, ad groups, or even locations. We recently analyzed a small budgets account targeting all of the U.S. In this instance, it will likely be a better use of your time to only keep the top 10-15 highest converting Geo-locations live, and then use that budget to purposefully, more aggressively test 1-2 additional cities or keywords.

<u>2) Invest in a smart burst-spend testing strategy</u>

Once you have freed up your budget, then focus on specific targets, locations, videos, audiences, or whatever is the next big thing you want to test in your account. It's crucial here to have a framework for what to try; otherwise, you may just waste money fiddling around with buttons.

- Do basic keyword research. What terms are people searching for that you're not currently targeting?
- Analyze cities or states or regions where your products are purchased through other channels. Is there opportunity for testing in locations you should be more aggressive in based on performance from other marketing efforts?
- Analyze competitor behavior. Are they invested aggressively in certain markets or audiences you aren't currently targeting?
- Is there an opportunity based on your content, to get more aggressive with a specific product in your Google Merchant Center feed?

Once you have an idea of what you want to test, you can simply take the money saved from the pullbacks in the rest of your account, add in a little extra for the smart test (even failed learnings are valuable!) and give it a whirl.

It may be easier to map this out, so let me illustrate in a fake account. Here's what it can look like.

An Example:
- Account: Sam's Barbershop (located in DFW, TX)
- Monthly Budget: $1,500/mo
- Search: 150 keywords live (all match types) spending $1,200/mo, avg CPC $2
- GDN: 1 campaign spending $300/mo, avg CPC $0.35
- Monthly Leads: 65 (avg CPL $23)
- Timeframe: P180D (past 180 days of data, keep in mind that SMB accounts often need far more time to build enough data before analyzing it with any certainty)
- Complaint: "I'd like to test additional keywords or figure out how to grow leads and our traffic, but we

just can't get over the plateau."

Here's where we would likely get into the account and then offer some suggestions (made up for the sake of this article, to walk you through how we think about this):

- o We eliminate the Google Display Network (GDN) for now (it is difficult to identify success for local businesses in the GDN, and we need the cash for more guaranteed demand capture opportunities). $300/mo saved.
- o We dig in and identify that 75% of their leads come from two geographical areas within the Dallas area. But those regions only account for 65% of the cost. We target only these regions for now. $420/mo saved.
- o We identify that of the 150 live keywords, only 60% of them have led to ANY conversions at all within our timeframe. We pause 60 keywords. They were low traffic, so $100/mo saved (general estimate for the sake of this article).
- o We identify that of the remaining 90 live keywords, 80% are within their target CPA within our timeframe. We therefore pause an additional 18 keywords. $80/mo saved.

Overall, we have now saved $900/mo WITHOUT LOSING ALMOST ANY TRACKABLE CONVERSIONS. We've trimmed the fat. Now here's the deal: admittedly we need to be careful of attribution and the way some of these keywords could be feeding customer leads through other channels, but… at this size of an account, that's a risk we have to take because the potential gains to be made utilizing that saved budget in new efforts could well account for what was lost.

Now, with $900/mo available, I'd recommend to a client that they invest an additional $500/mo for the next 2-3 months and we would put that newly found $1400/mo into some testing. Perhaps now we would aim at terms for which we hadn't had the budget before. Let's say the barbershop wanted to begin advertising shaves as well, and we could hit that hard. Or, perhaps there was a geographical-radius region (a local suburb) near their barber shop where the barber wanted to hit hard for a month with aggressive keywords, and GDN targeting including an offer for a 25% discount as a specific code on their landing page. Perhaps they want to make a play to get more aggressive in that area with brand awareness. They now have the opportunity to try these new ideas because they found a ton of budget without sacrificing the rest of their account.

This is an example of how to efficiently manage smaller budget PPC accounts and move them towards more specific testing.

Here's something crucial to keep in mind. In ultra-low budget accounts, you sometimes have to pause what could perhaps be working (you don't really know) but is simply low traffic, because the aggregate budget from those elements is more important to you currently than is the potential that those keywords will convert in the future. It's not that those keywords will never work, but it's that you're preventing yourself from possible testing opportunities for growth right now.

With small budgets, the knowable becomes far more important than the potential.

18 LET'S DE-FRENZY PPC

I have a secret to share.
I'm almost embarrassed by it in this hustle culture, but I'm also proud of it.

My secret is that I obsess over keeping client work at our little agency frenzy-free (as much as possible).

Sure, there's the occasional client mishap, last-minute promo, and late-night conversation (about twice a year for us, seriously). But, overall, nobody on my team works nights or weekends and we just don't really have many last-minute "urgent" requests by clients that require us to drop everything to service.

Let me repeat that:
We're a digital ad agency that doesn't work nights and weekends!

Hard to believe, huh?
The reason this is almost scandalous to admit is that PPC has become known as a frenzied activity. In fact, I've seen it worn like a badge of honor!

Unlike our professional counterparts, we must work night and day, managing budgets, replacing ad creative, and taking 2 a.m. calls from anxious clients. Why? Because, PPC! Hustle! Meaning in Life!

(Or so we're told.)
But, I'd like to push back on this culture (I know, big shocker

for anyone who has read my book Stop the Scale[23], or watched my TEDx Talk: Stop the Scale: Redefining Business Success[24]).

I'd like to make the bold claim that PPC doesn't have to be so frenzied, and for the good of everyone's mental wellness, we should begin making purposeful steps within our departments and agencies to de-frenzy PPC.

Rather than frenzy being a hallmark of PPC, I think we should work toward stability.

Stability is the opposite of frenzy and the goal for which to aim. Stability will improve your client relationships and employee morale (as well as your own mental well-being, if you're a PPC manager or owner).

With stability, you have the time and energy and resources (let's call that margin – more on that later) to actually focus on building and managing well rather than being constantly tossed around by the urgent nature of frenzied tasks.

So how does one push for stability in PPC? Shouldn't PPC, by its nature, be frenzied? I don't think so, and I will identify the seven key causes of PPC frenzy below – and solutions.

Ready? Let's begin to make healthy changes and de-frenzy our industry.

1. Mistakes

One of the primary causes for frenzy in PPC is the need to make up for a mistake that was made – oftentimes by the PPC manager, but sometimes by someone else (like the client).

For instance, a novice PPCer may be in a rush to hit their quota of tasks for the day, so they push forward a campaign with tROAS set to 20% rather than 200%.

They also set a daily budget far higher than it should be, causing the bidding algorithm to run amok in spending all of that client's monthly budget in one day.

---

[23] Kirk Williams, *Stop the Scale: Building a Digital Agency You Actually Like*, (Billings: ZATOWorks Publishing, 2022).
[24] Kirk Williams. "Stop the Scale: Redefining Business Success." *TEDx*, October 2023, https://www.youtube.com/watch?v=nqNBgMqpEzI&t=10s.

The client sees it that night and angrily calls the agency to fix it. Hurry! Fix this costly mistake! Emergency! Frenzy!

On one hand, it's an emergency because the agency needs to fix the mess it caused. On the other hand, it isn't a true emergency because it could have been easily avoided.

Not all mistakes are avoidable, but I think most are. In PPC, I actually think pretty darn close to all mistakes are avoidable.

The solution

Slow down and check your work. An ounce of prevention is worth a pound of cure, so build into your agency's process a plan for double-checking work as part of campaign builds and optimizations.

As an example of the kind of clever thinking you can build into your agency life to minimize mistakes, Collin Slattery (owner of Taikun Digital Agency) recently noted on Twitter this tip for budget management:

"All our budgets end in .99 or .01. These are directional indicators for budget changes. For example, if a campaign budget is $499.99, that means we most recently lowered the budget. If the campaign budget is $500.01, it means we most recently increased the budget. If the budget is $500.00, it means it is the original budget and has not been changed. This makes it incredibly easy for anyone looking at the account to know at a glance the directional changes that are being made to all our budgets."[25]

What a great example of the sort of process you can build into your team to make wiser decisions in places where mistakes are often made.

Also, make sure you're hiring people with strong attention to detail. I think PPC frenzy often happens because managers get sloppy, and I have found that hiring team members who notice things naturally is an immensely valuable soft skill to obtain.

Finally, if a team member has strong attention to detail but is making mistakes, it could be that the mistakes are coming from an unrealistic level of tasks and priorities.

Perhaps you need to rethink what your team can actually

---

[25] Collin Slattery. @CJSlattery. Twitter, 23 January 2024, https://twitter.com/CJSlattery/status/1749845234341605843?s=20.

accomplish within the allotted time they have in a day, rather than just assigning them all the things and expecting them to get it done. That is a key source of employee burnout.

2. Poor planning

Another primary cause for frenzy in PPC is the age-old failure of poor strategic planning. You've probably seen a version of this popular quote somewhere:

"Your lack of preparation does not my emergency make."

Well, the same goes for our own failure to plan in PPC.

A last-minute urgent email from a client leading to a team needing to work late in order to finish getting creative ads live before the promotion begins at 12 a.m. Eastern time isn't actually an emergency. The team should have had knowledge of the promotion for two months, and the failure here was in the last-minute nature of the request.

So what about when the client fails to adequately notify the PPC team of necessary changes?

This gets a little more complex, but communication and expectation-setting are important parts of PPC. In that case, an honest discussion should be had as to the correct way to notify of something like upcoming promotions.

Perhaps it could go like this:

"Hello, {Client}. We are unable to get that promotion live by the end of today since our team is heading out the door. However, we will jump on this first thing tomorrow. Also, let's plan to connect on our next call as to a good way to plan promotions together so we ensure these ads are live when they need to be for future sales. I want to make sure we do everything we can to set you up for success, and being included on your promotion calendar will give us the information we need to get this done on our end."

Or, you can always jump on the last minute request as a sign of good faith, and then use that as the springboard for the discussion on future boundaries and planning.

The solution

Recognize that emergencies based on a failure to plan accordingly can (and should) be avoided, and take the necessary steps to educate and train your team or collaborate with your clients to be able to anticipate the inevitable activity.

The great PPC team is the anticipatory team. Many false emergencies arise that could have been easily avoided had they simply been thought of beforehand.

For instance, if an important holiday is coming, and your team hasn't heard from a specific client on a sale, but your team knows this client loves to run promotions, then great planning means your account manager proactively contacts that particular client to inquire about a sale (and perhaps even introduce the thought to the busy client business owner) well before the week of the event – when the client suddenly decides they want to do a sale and causes everyone to scramble.

That is an unnecessary and poorly planned, non-emergency.

3. Disorganization

Ahhh, disorganization. This has caused many innocent-sounding account managers to feel more mentally stressed than they actually should be.

A disorganized account, task management solution or communication parameter leads to false emergencies.

For instance, when you land a client, there are certain things you will need to know about every single account that you take on to manage, and it's important to have an organized process for gathering, storing and accessing that information.

Failure to identify this, gather, store, and make accessible this data will increase the likelihood that this important information cannot be used for the good of an account.

False emergencies are worsened by disorganization as it leads to faulty decisions made in the account that would never have been made if the information to avoid the false emergency had been taken into account.

This is also where lack of margin (below) comes into play, and we'll discuss that at length. I have found that some people are naturally disorganized, true. However, some disorganization can come instead from poor internal processes and policies or too many tasks.

It's difficult for your account manager to be organized when they are stressed and overworked and jumping from task to task without time to think and plan ahead and organize their lives.

See how many of these things blend together?

The solution
Recognize that easily accessible and properly organized information lays the foundation for proper planning and proactive choices that stop false emergencies before they ever can get started.

4. Unrealistic expectations

While many of the frenzy causes I address can be applied to either client or PPCer, this is about the only one that is solely the fault of either a manager or client (i.e., not the PPCer). Sometimes, false emergencies thrust upon the PPCer are simply because the client or the PPCer's manager has faulty expectations for what can or should be accomplished.

Let's say a PPCer gets three voicemails over lunch and six emails from a client in the same timeframe. They are alarmed. "Call me back immediately!"

When the PPCer calls them, they hear something like, "I've been looking at our keywords and I noticed that CPCs have risen by 25% from yesterday! This is alarming, and something has to be done! I'd like to hear of three action steps you're going to take as soon as you hang up to address this."

In this made-up (yet not far off-base story), the client has suddenly taken it upon themselves to wipe clean the PPCer's task list for the afternoon to address what they consider an emergency but is not actually an emergency.

The reality is that those outside of PPC (especially those with some power over the PPCer) can have faulty views of what should occur. Thus, they can try to instill a sense of urgency onto the PPCer that may not actually result in improved account performance (the goal we're all after here!).

The solution

This may be one of the trickiest for the PPCer to navigate since this covers a whole lot more than simply planning better or getting more organized. It involves working with clients or managers who themselves may be viewing this the wrong way.

Unfortunately, since they hold the power in this dynamic, it can feel impossible to actually stop the endless "emergencies" that arise every time the client has a good idea and wants you to "hop on a call" to discuss it.

I think the solution here is at least two-fold.

First, it includes education and sometimes healthy pushback. "I got your message; can we plan to connect on this during our regularly scheduled call next week?"

If they say no, it's important (with your boss's permission) to push back a little with something like, "I apologize but I don't have the bandwidth today or tomorrow to discuss so I may have to wait on that idea until next week."

I will note that there are a million things that could come up here in the client's mind as emergencies and a million ways to reply, so growing in soft skills is the way to address this (not to only use what I say in this article).

The bigger picture solution here is more complex but even more important. That is to be able to identify signs of this sort of client before you ever land them (while in the discovery process), so you can consider whether this is a client to avoid ever taking on.

Or, you may determine you will take them on, but set extremely clear expectations as to communication cadence so you can stick to that in replies.

A great client, even an excited one who likes to hop on calls, will respect a tactful pushback because they properly understand the role you play in their growth.

5. Bad timing pushes

This cause probably fits under "poor planning" but I like calling it out because it's a really specific use case with a clearly achievable solution. This will be the shortest since it's so simple, yet rarely followed.

A significant number of emergencies arise from something big changing in an account. Therefore, never push anything big in late afternoons, or on a Friday. It's just not more complex than that.

The miracle campaign you have built that you really, really, really want to gather data over the weekend is just begging you to get pushed live on that Friday afternoon.

But, consider what would happen if something blows up with it and suddenly you're at the swim park on your phone trying to frantically lower bids in the Google Ads app while your kids splash around?

What if the campaign isn't actually a miracle? What if you need to monitor it more closely? Doesn't pushing it live on Monday morning make so much more sense?

Did the client want you to push it live before the Sunday holiday? Then plan better to push it live on Thursday so you have a day to analyze it so it's ready for Sunday.

Sure, there will be exceptions to this, but the goal is to knock out as many non-emergencies as possible so you have the margin (see next point) to deal with the actual emergencies.

Not pushing major initiatives live in the late afternoon/evening or on Fridays will prevent a lot of last-minute work when you're not technically at work.

Note, international campaigns are a different beast when it comes to time and management so apply the above accordingly.

The solution

Don't push stuff live unless you have the scheduled bandwidth on your team to be on hand for the inevitable monitoring and adjustments needed for big changes.

6. Lack of margin

This one is my absolute favorite because I think it's the most underrated, yet most important.

The simple fact of life is that poo happens. The coffee mug version is too sweary for my book, so I'll keep it to "poo" here.

One of the biggest things you can do to actually navigate true or false emergencies is to build time to deal with the inevitable into your agency structure, process and calendar.

Margin looks like a team that has the time and bandwidth to

deal with things, especially in key calendar times (think Black Friday week!) when the likelihood of something going haywire rises. It's not complicated. We just need to do it.

The solution

Here is what I have learned about margin over a decade of PPC agency management:

- *Actual emergencies (once the previous causes I discuss here are eliminated) are incredibly rare.* Like, really, incredibly rare. If you have hired the right clients, planned well, and built margin into your agency life, then an actual emergency may account for only 5% of what normal agencies consider "emergencies." Recognizing this itself and acting on it is a cheat code to stability!

- *Building margin into your agency means you can actually deal with actual emergencies.* For example, while our agency is naturally busier during Black Friday and Cyber Monday (BFCM) week, we purposefully build additional margin into the week to account for the unexpected by:
    - Canceling normal client calls scheduled for that week.
    - Slowing down or even pausing our normal account optimization schedule.
    - Spending more time in analysis and monitoring accounts for changes so we can be proactive in dealing with the unexpected. (2 a.m. work during BFCM tends to occur for other PPC agency employees because they have to fit the unexpected into their normal expected schedule. This happens very rarely for us, if ever.)

Literally, build margin into anticipated busy times for the unexpected, and any true emergencies will be able to be addressed in a way that doesn't actually sacrifice well-being.

Planned stability is a beautiful (achievable!) thing.

7. Actual, true emergencies (often platform-created)

If all of the previous frenzied causes have been addressed and planned for and organized, it is possible for the frenzy to be caused by a true emergency! Yes, there are actual emergencies that come up in PPC, and it would be negligent of me to not admit that.

I think one of the key ways we see true emergencies occur in PPC is when an ad platform makes an unexpected, sweeping change that causes disruption in our normal operations.

Google may release a new policy overnight that causes your clients' ads to get disapproved and you have to act urgently.

Google Merchant Center might disapprove a product for something random, and your client's top product disappears, so you need to clear your day to deal with this.

BFCM week is busy for ecommerce. It just is. Even with all of our team's planning, we still find that we work more on BFCM than a normal week with unexpected things that arise.

Another thing may be some sort of natural disaster or unrelated to PPC event.

A friend told me recently that he held a job a few years ago in which their CEO and COO went for a drive to check out his new sports car, lost control of the vehicle, and heartbreakingly, both of the key execs died in the event.

Horrible and unexpected events such as that can cause mental difficulties, margin stresses, and impact planning in ways that nobody could anticipate.

<u>The solution</u>

In many cases, simply admitting that emergencies can occur and building proper margin into key times can help assuage them.

As I noted earlier, our BFCM margin planning schedule is a way we've identified a time of the year when an "emergency" is more likely to happen and built margin around it.

That doesn't negate the fact that sometimes true emergencies happen, but eliminating the previous six causes of PPC frenzy will at least help give your team additional energy and mental acuity to navigate the actual emergency when it finally comes along.

They're not being thrown about from one emergency to

another until they finally burn out and leave with a bad taste in their mouths for your agency, or PPC in general. It just doesn't have to be this way!

Join me in de-frenzying your PPC agency, team, or personal schedule!

19 THE REMARKABLE AND ALARMING POWER OF THE GOOGLE ADS PLATFORM

Let's say you go to a gas station in a small town in Montana on Friday. I'm picking Montana since this is my home (now), and it's a more likely scenario of comparison than the thought of being limited to one gas station in the middle of a metropolis.

Anyway, you walk up to the attendant of said gas station and ask, "I need gas."

The attendant gives you an odd look.

You clarify, "for my car. I'd like to buy some gas but noticed I couldn't pay at the pump. Can I get a few gallons?"

He says, "Well sure, I have plenty of gas! Unfortunately, those two people over there also want gas and so I have decided you will bid against them for the gas in my tank."

"That's odd," you think to yourself. "But I really do need gas, so what choice do I have?"

You ask him what he needs from you.

"Well, here is how it works. You tell me how much money is in your budget for gas this month. Then I go talk to the others to hear how much their gas budget is, and then I decide what the winning price is, and award someone with the gas."

He goes on, "Also, you won't know what type of gas you got, whether premium or unleaded, and I won't tell you how much I put in your tank until you are leaving."

"Wait a second," you cry, "I'm not sure if I want to spend all my budget for gas based on what you determine I will pay for this fill-up, so can you just tell me what the price is for 2 gallons so I can find another town?"

He shrugs and continues with a slow drawl, "Nope, that's not the way this works. I run a private business, and that's my chosen business model. I have plenty of gas for everyone, but I make more money doing things this way."

He goes on, "oh and incidentally, there are no other stores that sell gas within 200 miles of here so if you want to use your car to leave town, you'll need to follow my business model rules... But again, you are perfectly free to choose not to use your car to leave town. It's a free market, after all."

This sounds far-fetched, doesn't it?

There are laws about things like price-gouging, transparency, truth in advertising, and a host of other things that would likely mean this business scenario I described could never actually happen in the consumer auto gas industry.

Yet, in advertising tech, we are looking at scenarios that play out in frightfully similar ways.

In fact, Google has arguably built a $70B business around this exact model.
- o How are the bids set?
- o Are there bid floors?
- o If so, what are they and what are they based on?
- o How can I determine if everything is above board?
- o How can I determine how my bids are set compared to competitors?
- o What is to determine Google isn't changing auctions behind the scenes for purposes other than what they've disclosed?

The unfortunate reality in 2024 is that we just don't have answers to most of these questions... which causes distrust within the very customers responsible for that $70B annual revenue.

We increasingly have some answers for these questions, of course. In a recent Federal antitrust case against Google, former Google executive, Jerry Dischler, admitted to auction manipulation technology such as RGSP, squashing, and changing bid floors around solely to hit Google quarterly revenue goals.[26] As I will point out in future chapters, the primary issue here to me isn't that Google changes its auctions (as long as it does so within legal constraints, of course), but it is that Google is communicating one thing and behaving another way ... and that ruins advertiser trust in an industry where trust is already spread thin.

I'd like to flesh out the power Google actually possesses in an auction environment a little more in this chapter, not to sow seeds of disruption, but to educate advertisers to the reality of their position, with the hope that this allows healthy advertiser pressure to be placed on Google for increased industry health.

The State of Paid Search Today

<u>Market Share</u>: Here is our current search as a channel scenario. When it comes to market share for Search Engines, Google owns 92.26% of the search market share worldwide.[27] I've heard less than this, but always greater than 75%.

For lack of a better way of putting it, this means if you want to advertise on a search engine, but don't want to use Google, your options shrink fast (by the way, Microsoft Ads is a great option to add as a supplement, but certainly no replacement for Google in sheer volume.)

<u>Advertising Revenue Share</u>: In 2022, Alphabet's revenues amounted to $279.8B, while advertising revenue accounted for a whopping 80% of that total ($224.4B across all Google ad channels).[28]

---

[26] Matt Southern, "Google Allegedly Adjusts Ad Auctions To Meet Revenue Goals," blog post, 2023, *Search Engine Journal,* accessed 2 February, 2024, https://www.searchenginejournal.com/google-allegedly-adjusts-ad-auctions-to-meet-revenue-goals/496634/.

[27] StatCounter, "Search Engine Market Share Worldwide," StatCounter Global Stats, accessed 2 February, 2024, https://gs.statcounter.com/search-engine-market-share.

[28] Statista, "Annual revenue of Alphabet from 2017 to 2023, by segment," Statista, accessed 2 February, 2024, https://www.statista.com/statistics/633651/alphabet-annual-global-revenue-by-segment/.

That means Google as a financial entity is larger than Greece or Ukraine, and more than double that of Guatemala… and that's predominately from its advertising platforms. This is not an insignificant size and should cause pause when considering the level of financial power Google can position in various applications.

Ponder some back-of-napkin math behind a lawsuit against Google, in light of their financial size. In June 2017, the EU fined Google a record $2.7B for "unfairly favoring some of its own services over those rivals."[29] This number is staggeringly large to us "normies" but when it comes to a company the size of Google, it's virtually a rounding error.

Just ten days ago from writing this, another update on the case landed in which a court advisor noted the ruling should be upheld. The case isn't even close to being finalized (all appeals exhausted, etc.) when Google has to actually cut a check to the EU for $2.7B since we're 6.5 years past the ruling already. In those six and half years, Google's annual revenue has grown from $111B[30] (2017) to an estimated $297B[31] (2023 estimated until official numbers are released in 2024 at time of writing this chapter).

This means, since the fine of $2.7B was announced, Google has made an estimated $728B in profit since that ruling.[32] Profit. We're

---

[29] Mark Scott, "Google Fined Record $2.7 Billion in E.U. Antitrust Ruling," article, 2023, *The New York Times,* accessed 2 February, 2024, https://www.nytimes.com/2017/06/27/technology/eu-google-fine.html.

[30] U.S. Securities and Exchange Commission, "Alphabet Announces Fourth Quarter and Fiscal Year 2017 Results," SEC.gov, accessed 2 February, 2024,
https://www.sec.gov/Archives/edgar/data/1652044/0001652044180000 04/googexhibit991q42017.htm#:~:text=%22Our%20business%20is%20 driving%20great,up%2024%25%20year%20on%20year.

[31] MacroTrends, "Alphabet Revenue 2010-2023 | GOOG," Macrotrends LLC, accessed 2 February, 2024,
https://www.macrotrends.net/stocks/charts/GOOG/alphabet/revenue #:~:text=Alphabet%20revenue%20for%20the%20twelve,a%209.78%25 %20increase%20from%202021.

[32] MacroTrends, "Alphabet Operating Margin 2010-2023 | GOOGL," Macrotrends LLC, accessed 2 February, 2024,
https://www.macrotrends.net/stocks/charts/GOOGL/alphabet/operati ng-
margin#:~:text=The%20current%20operating%20profit%20margin,30%

likely years away from them having to pay, so it's possible Google will have made $1T in profit, not revenue, profit by the time they need to even pay that fine. $2.7B doesn't seem so much now, does it?[33]

The point I am making here is that Google is of the size in which historically significant governmental fines hardly make a dent in their operating expenses and margin. An entity that size should be carefully considered as to the power they believe they wield.

Advertiser Support: Recently, Google laid off another round of support representatives,[34] even while they (1) recently set in motion a paid support service requirement,[35] while (2) advertiser complaints about support are (from my anecdotal position) at an all-time high.

In lieu of this, I recently joked on Twitter/X, with only a little irony:

"How to know you're a monopoly #476: Your customers who pay you to use your service have to pay to get support to use your service and then you lay off a bunch of those support people anyway cause lulz what are your customers actually going to do and where would they go?"[36]

2C%202023%20is%2022.46%25.

[33] I think it's important to note that the numbers I toss out here are rounded, and calculated based on publicly provided sources of data. It's possible these are not entirely accurate, though I think the main point I am making remains, which is that Google will make a heck-ton of profit before they ever have to pay back any fine... which suggests the fine will not have its intended affect.

[34] Nicola Agius, "Google Ads lays off hundreds of staff amid support crisis," article, 2024, *Search Engine Land,* accessed 2 February, 2024, https://searchengineland.com/google-ads-lays-off-staff-support-crisis-436589.

[35] Barry Schwartz, "Google Ads Tests Paid Customer Support Options," article, 2023, *Search Engine Roundtable,* accessed 2 February, 2024, https://www.seroundtable.com/google-ads-paid-customer-support-options-35897.html.

[36] Kirk Williams. @PPCKirk. Twitter, 19 January 2024, https://twitter.com/PPCKirk/status/1748378443757052300?s=20.

Smart Bidding Control: Google is maneuvering to fully control bidding with its Smart Bidding solutions.

This means they control how much each person in an auction pays to... well, themselves.

Let me say that again: Google controls the bidding among separate parties to determine how much they themselves get paid. At the very least, the advertiser can see what they pay after the fact, but how much did an advertiser need to pay in order to get that ad rank? That is hidden by...

Ad Rank: Google utilizes a metric hidden from external analysis that ensures changes can be made within the bidding system to manipulate the bid without any third-party oversight.

This metric is called Ad Rank.

With Ad Rank, Google possesses the private means to make a bid whatever they want that bid to be, with no external verification as to whether the final outcome is authentically priced according to the public-facing pricing motivators Google has led the advertiser to believe are occuring.

While they reveal some details through the external-facing, Quality Score metric, it only reports a score between 1-10 and includes assumptive figures like "expected Click-Through-Rate" (again, a made-up Google metric that has not actually occurred, it is based on predictive estimates) as an unknown percentage of the Score.

Put differently (and somewhat bluntly): if Google ever wants to commit fraud within its system (and I am not here suggesting they have done so, just revealing the mechanism in which fraud could easily occur), the tool is there to hide it within plain view since they could easily manipulate Ad Rank for a specific company in an auction with no one the wiser.

Data Obfuscation: Google continues to remove data that allows for third-party authentication into the quality of auctions.

Whether for legitimate privacy reasons or not, the recent search terms decision[37] to no longer show search terms without "a

---

[37] Susan Wenograd, "Google Ads to Start Hiding Some Search Query Data," blog post, 2020, *Search Engine Journal,* accessed 2 February, 2024, https://www.searchenginejournal.com/google-ads-to-start-hiding-some-

significant amount of data" is simply one in a long list of automation decisions made to purposefully obfuscate what is happening in targeting.

As I have written elsewhere on Smart Shopping, and then Performance Max, an advertiser or brand can no longer verify the keywords or audiences were acceptable targets because the data is no longer there. This lack of data also prevents other helpful uses, such as utilizing search term reports for SEO or Google Shopping title ideas.

The fear of "control" and lack of data-reported back in accounts by PPCers (which often causes eye-rolling by Googlers), is about more than self-preservation; it is the need to see into the workings of Google to ensure advertiser money is being well-spent. It's healthy for us PPCers to seek transparency since we are striving to be good stewards of that money, and why would we ever trust the auctioneer in this analogy I described above?

Thankfully, Google has begun to offer additional insights such as Search Terms insights into Performance Max campaigns, though one wonders what other search terms are being matched to that Google did not determine should be revealed to the advertiser. But it is not wise for Google to push back on the friction that necessary transparency carries with it. Transparency does slow down innovation, but that's part of the ecosystem they've chosen to engage in.

With these things in mind, it is clear that the current trajectory of the Google Ads system is one that positions Google to maintain sole control of all aspects of the advertiser auctions with little useful data fed back to advertisers.

But wait, there's more.

Let's talk about the cookie.

We are in a curious dilemma currently with the cookie. On one hand, I applaud the desire of governments to protect user privacy and help champion user data rights by limiting who has access to user data. I think there is a lot of good happening as these conversations take place.

However, as the war of the third-party cookie occurs, I am

---

search-query-data/379840/

concerned about an ill-defined side effect. In the world of 2020 technology, data is power and literally at times is currency. Data is worth its weight in gold, and as third-party cookies are removed, the only advertising entities left standing with that weighty gold of user data are first-party cookies placed most often by platforms such as Facebook and Google.

The war on third-party cookies is supposed to be limiting advertiser access to user data. Yet, it is arguably increasing the ultimate power of the advertising platforms. Now we have a scenario in which entities like Google control not only every aspect of the auction, as I have outlined above, but also have increased access to data not available to others.

In one federal hearing room, you have legislators seeking to limit the power of the platforms as they pursue antitrust cases. In another room, you have them handing them exceptional power by ensuring – through their first-party cookies – that they are the last ones standing.

> To be clear, I'm not against third-party cookie limitation.
> I'm not anti-Google.
> I'm not anti-tech.

I actually write posts like this for the very opposite purpose than the concerned tone would suggest.

I love the paid search industry I am in and hope to see it and Google (as well as other platforms!) thrive for years and years to come.

I do not believe thriving will occur if obfuscation and decreased transparency continue. If we believe that a company the size of Google, with its increasingly black box transparency, will not eventually encourage fraudulent activity, then we haven't read enough history.

I don't know the best way to fix this in totality, but I do know that it would be a great start for Google to step back from limiting advertiser data access and to begin treating its advertisers in more of a partnership manner to encourage healthy transparency and long-term growth.

As always, may the auctions be ever in your favor.

APPENDIX A

To my friends at Google,

It is worth beginning by noting that I do not represent the diverse PPC community as a whole, but this letter will hopefully demonstrate some representation of concerns shared by other members of the PPC Advertising community regarding ongoing practices within Google that I believe have overall caused more harm than good to users, advertisers, and Google itself.

That being said, I think it would be remiss not to acknowledge my appreciation for Google and this industry. While this letter cannot, and will not, speak to any legal or ethical matters in regard to any past, current, or future cases brought against Google, I think it fair to acknowledge that Google has grown in its popularity in large part due to its high quality products. The innovation, creativity, and practicality of Google products are enjoyed by millions (billions?) of people worldwide, including ourselves.

The author of this letter values Google and its contribution to the world, and it is from this standpoint that I seek to write this plea.

My desire as a PPCer is for the good of my industry, since my livelihood and career depends on the success of Google. This letter is not intended to be an attack from an external force, but a plea for self-examination by those within the industry I care about so deeply.

I also value the people who compose Google's nearly 140,000 person payroll, and this letter is not intended to disparage any single individual and their contribution to our industry. It has largely been the author's experience that Googlers are a people who care deeply about leaving a positive mark on the world and doing the best they can to help businesses through Google's ad products. I acknowledge this, and am thankful for these individuals.

Decrease in Advertiser Trust

With this in mind, I would like to address the overriding purpose for this letter: to present a list of concerns to Google (as a singular entity) which I hope will highlight the trust I believe you have lost with the advertising community and will continue to lose without a change in direction and behavior.

While the advertising community is never short on complaints regarding platform behavior, I believe there looms a larger, more ominous threat to the industry that revolves around the relationship between advertisers and Google, specifically in regard to the trust Google continues to hinder. I believe this trust will continue to degrade without a change in behavior by Google and will eventually impact the overall health of the industry as future innovation undoubtedly offers alternative solutions.

It is this trust for which I hope to make a case, as well as offer suggestions within the remainder of this letter.

Stated Concerns

There are at least six primary trends that have arisen to harm advertiser trust with Google over the years.
1. Google's place in the overall ecosystem as buyer/seller/auctioneer - While I acknowledge the complexity of the Google auction system, the amount of control Google exhibits decreases trust when paired with the lack of third party oversight into the inner workings of

the auction environment.
  a. Any industry would have concerned entities in a scenario where one party controls the tracking, auction, buying, selling, bidding, and then receives the proceeds from that entire process.
  b. While this level of control does not itself constitute improper behavior, it certainly leaves the advertisers in a risk averse mindset, since the only option advertisers have in that scenario is to completely avoid the auction itself... a difficult decision to make when Google is one of the only primary available digital options online for advertising and the most significant Paid Search option by a wide margin.
2. The continued product development moving towards Google-controlled automation - While I acknowledge the benefit of machine learning in the buying process and its remarkable capabilities, advertisers also recognise the reality that more automated campaign types within the Google Ads system means there are more opportunities for Google to make changes within the auction system that would be disadvantageous to the advertiser, and advantageous to Google, without any third party knowledge.

3. The decrease in reported data - While I acknowledge that some level of privacy and security is necessary in proprietary auction mechanics to prevent bad actors in the advertising world from taking advantage of the transparency to game the auction system, advertisers also experience frustration by the continued decrease in reported data back to the advertisers.
  a. For example, the detailed search term reports (including helpful KPIs such as Cost data) which have disappeared from Smart Shopping and then Performance Max campaigns completely, have begun to be limited in Search campaigns. The only way advertisers can know the auctions in which to invest more aggressively (or use that information to benefit other marketing efforts such as SEO or Landing Page optimization) is if we see which

auctions performed. It is challenging for advertisers to request additional budget from key decision makers if 25% or more of the search terms are hidden.

  b. One thing is clear: automated campaigns that allow Google to control almost everything within the account and then report back on very little in terms of what is happening in those auctions erodes rather than supports trust. Rolling these automations changes back, or simply slowing them down, would likely strengthen advertiser trust.

4. <u>Lack of consistency between public and private admissions of what occurs in the auctions</u> - While I acknowledge there are hidden mechanisms to the auction that advertisers do not (and likely cannot fully) understand, advertisers also experience a significant hit to trust when hearing contradictory statements presented by Google representatives in the way their auctions work.

  a. As one recent poignant example of this point, the VP/GM of Ads at Google, Jerry Dischler, responded to questions in 2015 at the SMX Conference[38] and he denied that Google was making any changes to the auctions for revenue building purposes. Then in 2023 during his testimony in front of the federal committee, Dischler himself went into detail regarding the specific mechanisms[39] Google has employed to

---

[38] Andy Taylor. @PronouncedAhndy. "Jerry Dischler in 2015: 'Full stop, we're not manipulating search results or manipulating the ad auction to increase profit. That's just not what we do.' (check the video link: https://www.youtube.com/watch?v=0m6KINWDEqs) Jerry Dischler in 2023: 'we tend not to tell advertisers about pricing changes'." Twitter, 21 September 2023, https://twitter.com/PronouncedAhndy/status/1704861673155080362?s=20

[39] Nicola Agius, "Google denies manipulating ad auctions in resurfaced SMX Advanced clip," article, 2023, *Search Engine Land,* accessed 2 February, 2024, https://searchengineland.com/google-denies-manipulating-ad-auctions-in-resurfaced-smx-advanced-clip-432345.

purposefully build additional revenue into the auctions at key times in order to hit Wall Street projections. These include RGSP, reserve price floor surges, and Squashing.

    b. The point of this letter is not to determine whether Google has broken any legal or moral obligations; it is simply to plead with Google that this behavior of contradictory public claims (as well as public claims misaligned with private practice) harms advertiser trust.

5. <u>The impact on product quality on users when revenue goals become more important than innovation and quality</u> - While I acknowledge that every business has the right to pursue its revenue creation goals in a healthy manner, advertisers recognize the concern that this pursuit not harm product quality.

    a. In an email conversation shared during a recent DOJ hearing,[40] it was revealed that the Ads (PPC) side of Google was pushing for the Organic (SEO) side of Google to make key changes to a roll-out that had already occurred, simply to increase Google revenues. To his credit, Anil Sabharwal (Product VP and GM at Google) pushed back on this request by the Ads department, and I do not know the outcome of this conversation.

    b. However, surely I can agree that it harms advertiser trust to become aware of a request to roll back a product launch that had already gone through Google's rigorous standards tests for a high quality user experience, all in order to push for more revenue.

    c. This is a clear instance of at least one Google executive accepting the cost of a user product

---

[40] Trial Exhibit. "UPX0522: U.S. and Plaintiff States v. Google LLC." Justice.gov, 2023, accessed 2 February 2024.
https://www.justice.gov/d9/2023-09/416646.pdf?fbclid=IwAR1CTqklvMuFZg11V0Qt01pg2w_bXM3P20wUlhqxNKB3mTUSKs9alKqfuzg.

quality decrease for his primary goal of pursuing revenue. I would be naive to think this was the only case, and this process revelation causes concerns by advertisers as to whether Google product quality has been reduced over the past few years for the sake of hitting revenue projections.

6. <u>The lack of true support from the Brand/Agency Partners program</u> - I acknowledge that Google has no requirement to offer advertising support as the primary relationship between advertisers and Google.

   a. In an open letter discussing trust, I would be remiss in not offering the suggestion that a Google Partner program that appears to be about support, but is in actuality a sales program, erodes rather than supports trust.
   b. There is a heavy awareness within the advertiser community that Google representatives will be tempted to pursue their own financial objectives (Google Rep commissions based off of things like feature adoption) rather than primarily the advertiser's best interests.
   c. Admittedly, there are many good people within the Google Partner support program, but I suggest the entire system is geared towards supporting those who are better at sales than support. This is represented in the many contradictory facts received by various advertisers through different reps and shared online. It would build trust for advertisers to actually get the help they need and desire, rather than for Google to see this as primarily a new revenue opportunity.

Proposed Solutions

With these in mind, I wanted to propose suggested resolutions to help rebuild advertiser trust.

I want to first admit that I do not think the solution here is to require all proprietary algorithms to be fully transparent, as I understand the desire to prevent bad actors on the advertiser side to game this system in a way that would harm others.

I also understand that I have limitations to my understanding of how to resolve these concerns, having no knowledge of the internal product intricacies and future roadmap.

Therefore, I will outline a few suggestions, but leave this ultimately as an open-ended question to Google.

- Since there is a default from advertisers of lack of trust based on the behavioral points outlined above, what will Google do to demonstrate trustworthiness to your advertisers?
- How can brands be assured their billions of dollars will be handled in an ethical way that pursues positive and fair conditions for all (including Google) in the marketplace of the Google auctions?
- What can you do to rebuild trust with advertisers?

These are the sort of questions I do not think I as an advertiser can answer, but I would urge you to begin asking more directly within every meeting and get-together within Google.

Initial suggestions for strengthening trust with the PPC advertising community:

1. <u>Increase transparency in auction mechanics and cease activity that unbiased entities would consider illegal or unethical.</u> For instance, if you would like to build in a 5-15% price floor threshold to increase Google revenues in a down quarter, don't state that you are not engaging in this behavior. Rather, please be upfront with advertisers on these CPC changes. I understand that the cost of getting access to the amazing innovations that Google puts out is Ad Revenue. At the very least, understanding when to anticipate Platform-created shifts in average CPCs will help advertisers budget more successfully.
2. <u>Pursue product quality for Google search user experience over revenues.</u> I believe the primary benefit to Google has always been that it is an excellent product for its end users. I would suggest refocusing on product quality over pursuing revenue growth.

3. **Increase reported data to advertisers and allow for more advertiser's control in automated campaigns**... even if that at times means the advertiser account may not perform as well as Google engineers believe it can. For the reasons stated above, I believe this is a necessary step for advertiser user trust to be strengthened due to the level of control Google maintains in every part of the auction, especially as I have found that advertising and marketing are equal parts art and science. Automation can at times lean too heavily into the science aspect, believing all things can be determined by an algorithm with an expectable output... which is not always the case.
4. **Transition the current Google Partner Rep program into a support system.** I would strongly suggest pursuing a support-first approach to advertiser communication to help strengthen trust. You may be surprised how you are able to accomplish your revenue goals with this model as advertisers learn to trust their rep suggestions, become further educated in the skill of PPC advertising, increase advertising success in the accounts, and spend more with Google. This is a scenario in which everyone will win.

    I hope you will consider these things. It is with hope and enthusiasm I look to the future. I am optimistic that advertiser trust can be rebuilt with the right changes since I believe in the value of Google's products and its contribution to the business world.

    I believe that, with the right changes to increase advertiser trust, the PPC industry and community can continue to grow in success and strength for decades to come.

Thank you for your time,

Kirk Williams
ZATO PPC Marketing
December 2023

Ponderings of a PPC Professional: Revised & Expanded

APPENDIX B

What would a book on PPC philosophy be without a chapter on pricing PPC services? A pretty boring book, that's what! Here is a revised and updated version of the chapter included in my first book. Enjoy!

Getting your pricing right is important. If you mess up your pricing, you either can't sell because you are priced too high, or you can't survive basic business operations and your own mental health because you are priced too low. The problem is that there are a number of ways people look at pricing PPC, but there really is no "normalized" way of doing so, as you may observe in other professions such as accounting.

Let's be clear about something right from the beginning of this chapter: there is no perfect pricing model. I may have to repeat myself on that a few times to make sure it sinks in, because I will argue for a specific model as my preferred model. However, I maintain there to be no perfect model, and in fact, different situations may call for different pricing models.

I would expect a 250-person agency to think differently about pricing than a solo consultant, but that doesn't mean there is a right or wrong model that should be applied to every situation. That being said, I believe each of the pricing models contain basic qualities that will make them more or less suitable in pricing for PPC.

I also believe that every single model has misaligned incentives, and I'll go into that in each of the sections below. One of the common attacks on the % of spend model is its supposedly misaligned incentives, but that does injustice to the reality that every model has unique incentives between the vendor and client. There can certainly be incentives that align better than others, but let's not pretend models like % of revenue actually carry aligned incentives. This naive belief sets a brand up for being taken advantage of since the true incentive of the vendor is hidden from them.

I also believe that because there is no perfect model, it's more important to hire the team or person than the pricing model. The reality is that if you (a client) find the vendor you believe has the best overall strategy for your account and has a price with which you can work with, then who cares about their pricing model. Work towards an agreement that satisfies both of you, as once both of you are happy, the real work can begin.

Ready to roll? Let's dive in!

To discuss these, I want to focus on the primary PPC pricing models in the market today. Most of these have not changed in years; they are simply the way they are. We will discuss the pros and cons of each model and think through the "most ideal" pricing model.

Hourly

Hourly pricing is common for the solo consultant, in part, because this is how we have been trained to think. We see our value as directly related to our time. This is also a benefit of hourly pricing, because it puts our pricing into the language of many potential clients. We can all wrap our minds around, and make projections for, hourly rates.

Another benefit of hourly for the client is that they know someone is actually working in the account. Someone is "putting the time in" and they bill for actual time worked, not simply a monthly retainer.

However, that being said, if someone can avoid working in a monthly retainer setting, can't they also be less than honest on their timesheet? I don't personally buy this hourly rate argument, since people can lie about whatever they want to lie about.

Transparently, I'm just not a fan of hourly rates, as they carry too many problems for the vendor. With hourly pricing, you will find that you punish yourself for efficiency. Figured out a way to cut your time in half with a spreadsheet on a task? Congrats, but you also just cut your pay in half.

The same principle applies to finding new tools to help manage the account. If you decide to begin using tools to help automate a process, then you have just shot yourself in the foot by reducing your pay (and you still have to pay for the tool on top of it).

Of course, some may say that you should find other work to do now that you have been freed up. There is some truth in that, but not the whole truth. After all, even though you were contracted based on time, there is this troublesome thing called "performance" that you will have to deal with eventually. You could work yourself to the bone, maximizing every hour you have, and still end up losing an account if it doesn't send sales. I see this as a classic case of payment/deliverable mismatch. You are being paid for your time, but what the client actually cares about is the quality of your work in the account. If you actually increase the quality of your work, your pay does not increase, and vice versa.

Here is where the misaligned incentives lie: the client is most interested in performance, whereas in an hourly model, the agency is incentivized by taking longer to do tasks. I'm not saying all agencies will do that (again, hire the agency, not the fee model), but that the incentive isn't aligned with the client's.

As you can tell, for all those reasons, I'm not a fan of hourly except in the case of freelancer or short-term consultants. In my agency, we charge hourly pricing when we are contracted for training, or in-person consulting, since we have calculated a price based on our knowledge and time. I think this can also make sense for a freelancer who whitelabels for an agency or brand, since there can be more stable tasks they are taking action on which can line up well with specific hours.

Project / Task

Another pricing model that fits outside the realm of the monthly retainer models (we'll get into those next) is the project or task-based pricing model. This is also typically seen in consultants, rather than established agencies, though it can be seen as add-ons into an engagement if the scope of work expands. I most often see this model in PPC around account audits. They are short-term engagements with a specific deliverable (the audit); once the audit is completed, the relationship is over.

I am more attracted to the Project or Task based pricing model because it is built around accomplishing a specific objective. If you finish your task in half the time you assumed, you still get paid the full amount. Though even then, some calculate their projects based on an hourly rate so time is not totally lacking in the equation here. This is a reasonable way to come to a project rate, especially if it is on a project with some normal level of stability to it so the vendor is not being "punished" by finding a quicker way to do a task. In that case, they can simply increase their project pricing for the next audit. As with anything, keeping an eye on market pricing is valuable if you are able to keep your ear to the ground to determine what is a "common" price for the service you are offering.

Overall, I only have two main complaints with this model, but they are big ones.

First, this is the sell-till-you-drop model. If you are a consultant, you have to be selling constantly. Once you finish your project, it's time to find another one, and another, and another. This sort of work can get exhausting and (except in rare cases of someone being excellent at sales) appears to work the best within the framework of some sort of relationship with a constant supply of tasks (Fiverr or Upwork comes to mind).

Second, like hourly, it does not compensate according to skill. You could do a terrifically terrible job on that audit and still get paid for it. Or, you could identify the one thing to fix in an entire account spending millions and get compensated the same as if you had never found that opportunity, but still finished your project.

Because of these two elements, I am not personally a huge fan of project-based pricing.

Let's now move into models that have a recurring aspect to them. I will reveal ahead of time that I am a big fan of recurring sources of revenue as a vendor. We have had clients for years at ZATO where we get paid monthly, every month. I do not have to constantly sell to replace those clients, so they are valuable to us, just as we have proven our value to them. I am incentivized to maintain that relationship by doing a great job since maintaining that relationship is in my financial best interest as a vendor. Recurring revenue is the bees knees, but there is a lot of argument over the best PPC pricing model that recurs monthly, so let's discuss the different models.

Monthly: Performance

The Performance model is a model that attracts many, and maims those it wooed. That's a little unfair, but only a little. Note, this model is making a bit of a recurrence lately, so I spent a bit of time on it in this chapter. I often hear this trumpeted online, usually by brand owners, as a welcomed model. "Our incentives are aligned!" they'll shout, gleefully.

I first started freelancing using a performance-based percentage as my model because it seemed to be the elusive, perfect PPC pricing model, since agency and client goals were (so I thought) aligned.

The business owner is happy with the growth, I get paid more, so what could go wrong?! I quickly learned there were some gaping holes with this pricing and I ended up ditching it. Below are my primary issues with profit-based performance for PPC fees. As a quick note, I am talking primarily about profit pricing rather than gross revenue pricing. This is because any client worth their Excel skills will realize that profit is key to Ecommerce, not revenue. Revenue means very little if money is not being made at the end of the day, so I am focusing primarily on the profit-based pricing model.

Should you calculate in-channel or actual revenue/profit? A variance to consider in this model is whether you should calculate your fees based solely on the channel you (the vendor) manage or total revenue/profit. There are issues with both, and we'll discuss those more in detail next.

Who is truly responsible for the sales? Profit-shared fees on store-wide revenue assume the primary purpose of PPC is to make sales, whereas the true purpose of PPC is to identify and send people most likely to convert at different stages of the buying funnel. Read that sentence again. Even though it may not seem like much, this is a pretty crucial difference in understanding.

The reason this is important is because (most) PPC agencies do not control all aspects of the sales funnel (see previous chapters in this book making this case). Sure, we can give CRO advice, and heck, even control a Landing Page. However, that is completely different than helping run all aspects of a business involved in creating a truly desirable product/service priced correctly in a fluctuating market, training a sales staff (who can close sales), and ensuring all aspects of the online presence are primed for sales.

When PPC fees are tied directly to the store profit, this acts as if the PPC agency controls more than it actually does. As one (it really happened to me) example, what happens when the dev team pushes an update to the website on a Friday afternoon, preventing people from checking out for an entire weekend. PPC ads are still running and spend is still accruing, but revenue is zero. With a profit-shared model, that would work against the PPC agency for a client issue that had nothing to do with the PPCer. Another example is in a lead gen company for which we managed ads. We hit and exceeded their goals of sending phone call leads, but they had such call-center issues that they consistently had hang-ups because the phone was not being answered. We were doing our part (sending phone calls), but they were not closing them. Both are crucial to success.

How do you accurately track attribution in this model? Another weakness for me with this model is that it oversimplifies attribution and incentivizes the agency to make decisions which could negatively impact the top of the user funnel (where tracked sales are less likely to occur) when fees are calculated for in-channel tracked revenue/profit. Marketing channels work together in far more complex ways than we ever used to imagine (see the previous chapter on the marketing funnel), and basing pay upon a single channel is shortsighted and irresponsible.

There is still a great deal of guesswork and fuzzy math in models for which click should get what percentage of credit. It's especially problematic here, because now we are forcing the fuzzy math to be artificially clarified for the sake of our fee profit calculation. Practically, the greater concern is that the agency has to fight the temptation to work against the rest of the marketing channels when the agency fee is tied to profit. Bottom of user funnel traffic is a classic use case here, especially, for instance, if the agency would lean into remarketing and brand traffic heavily here.

With remarketing and brand, we PPCers are often only confirming a person's desire to purchase, even though they would have likely purchased through some other channel anyway. In this example, the agency unfairly increased their fees by purposefully investing the client's limited budget in a place that wouldn't truly grow sales incrementally. They only ensured all of those (guaranteed) sales were tracked as PPC in analytics, so their fees could be higher.

How do you get past the data tracking issues? Finally, the last reason I am hesitant to use an in-channel tracked profit model of PPC pricing is the data component. Data has interesting aspects about it in this regard. Is a CRM, or third party system such as Google Analytics being used? What happens when tracking issues are discovered? We just worked with a client to fix an issue we found in their Google Analytics Shopping Ads data. The GCLID was being dropped, and the Shopping Ads revenue was being underreported by up to 50%. If I had a percentage of profit model with this client, what should happen next? Should I demand the fees for the period before this was discovered and fixed since it was not my issue? This is not a small business, but a smart, recognizable brand.

Data issues happen all of the time, and tracking profit and trusting the data (along with the previous attribution issues) is yet another issue here that can complicate things. Clearly, I'm not a fan of the performance model. The only way I would consider it is how I have heard others describe their usage: they combine two models, such as flat rate and performance, to ensure there is a base pay, as well as an add-on "bonus". I still believe my concerns above would make this difficult and complex, and I find that clients value simplicity (as do I). During the sales process, I want to focus our energy on the value we can bring to their account, not wind up in endless conversations about our pricing model. Either way, I avoid this model.

Monthly: Flat Rate

With good reason, the Flat Rate model is a popular one these days. It allows for an agency to estimate the resources, time, tools, etc. needed to manage a PPC account and give the client a hard number for what it would cost to manage their Google Ads or Microsoft Ads account. It is nice for the client because they can get an idea of exactly what to set aside for their budget projections and keep things sewn up nice and tight.

I really don't have too many negative things to say about this model. The one gripe I have with this model is that it is not ideal for managing scope increase or account success. It still carries with it the same weakness plaguing our previous models, which is lack of potential income growth based on skillful account management. In that way, I think the incentives are misaligned. It doesn't really matter if the vendor does a terrible or a great job at managing an account, as they'll just keep making the same amount. Eventually, a client can fire the vendor, of course, but if we're talking strictly about incentives, then the two are not aligned. That's okay, since (as I'll repeat ad nauseam) no pricing models perfectly align incentives between client and vendor, but it's worth pointing to the specific variance in the incentives in this case.

Often in a PPC account (but not always), additional spend investment will result in additional scope by the vendor (as well as a variety of other requests from the client), but with the Flat Rate model all additional scope growth means renegotiating at the sales table. Call me sensitive, but I hate going back to the sales table with a sale I have already landed. I believe being back at the sales table can send a message to shop around and reconsider options. "Hey, if we're renegotiating again with our current agency, let's see what else is out there." If you do a great job in the account and triple your client's revenue and spend, you get a pat on the back, maybe. If you want to be rewarded for your skillful account management with the Flat Rate model, it means renegotiating, which puts you back at the sales table. I'm not a fan, but we're getting closer.

Monthly: Percentage of Spend

The most common pricing model has been (and likely, still is) the Percentage of Spend model. In this model, the agency bases its fees on the amount of advertising spend the agency directly manages. This can be similar to a flat rate fee, in that the agency calculates a certain level of profit they need to make from their fees, and prices their percentage accordingly. Interestingly enough, this harkens back to the Madison Avenue advertising days of traditional media, and some PPCers dislike it because they believe it is an outdated method.

The pros of percentage of spend are many. It is a fairly simplistic model to sell, it is common enough that it does not take a lot of sales effort to explain it (the prospect is often already familiar with the concept), but most importantly, it is scalable. In other words, it allows for the agency to be rewarded for a growing account. Of course, this is where the most vocal opponents will fix their sights. They will claim that an agency is in charge of managing ad spend, and yet, makes money when that ad spend grows... a clear conflict of interest!

In other words, "the incentives appear to be misaligned!" This is a legitimate concern, and undoubtedly with untrustworthy agencies, one that has proven itself correct. However, if you remember our oft-repeated mantra in this appendix, any pricing model can be gamed by dishonesty, and all pricing models have misaligned incentives.
- o An hourly worker can lie about his hours.
- o A flat rate agency can charge a fee and never work in the account.
- o A percentage of profit agency can manipulate data or ignore top of funnel users (stunting overall account growth) just to make their results look more impressive

than they are.

Any model can be gamed, which weakens this accusation. The response to this concern is fairly simple. It is one in which the agency never raises spend above a client-assigned monthly budget; and if spend rises past the assigned monthly budget, then the client is not responsible to pay the additional fees from the non-permitted spend increase.

This is how ZATO manages spend and encourages client trust, and it has worked exceptionally well for us in this regard. If we believe there is a case for increasing the budget, we will bring it in writing to the client and only raise the budget once the client has signed off on the approval request. Another concern with the percentage of spend model is that accounts can increase in scale and size without necessarily being a direct result of the advertiser work (for instance, in seasonal times such as the Christmas holidays).

There is truth to this concern, but there are also other factors to consider. The percentage of spend model (or any model based on spend) is simply providing a measured way from a third party verifiable entity (such as spend) to assign parameters around a measured spend growth. In other words, an agency can argue quite well here that this is simply the pricing model we have chosen to utilize, and a prospect is free to accept or reject it. If I don't want to pay extra money for premium gasoline for my car, I can shrug my shoulders and choose to pay for the bottom tier. I don't need to understand the exact details of why low-grade costs one amount and premium costs another. I don't mean to be insensitive here, but, at some level, an agency is its own separate business and doesn't need to apologize for choosing to price a certain way (as long as everything is ethically above board, of course).

The second part of this response, however, is that a spend increase often does come with additional scope (and risk) for an agency! More spend means more data and more optimizations to make more quickly. Do you know why the Christmas holiday season is so crazy for agencies? Because with the spend spikes come increased focus on setting up and running promotions, monitoring and jumping more quickly onto troubled campaigns and accounts, and a host of additional work that comes with

November and December. We get paid more as an agency in November, but that is also the month we work the hardest. We see this happen in normal spend growth as well. We have clients where we have doubled spend in a year, and that means we are always doing more work in those accounts than the year before. If you are getting data twice as fast, you need to be making optimizations twice as fast, which means more work. It doesn't always equate to the same amount of work, but that gets back to the things we have discussed previously on pricing according to return and to skill, as well as to workload. You may have just done a fantastic job and increased profits. Hey, if the client is delighted and you got an extra bonus based on your normal pricing model, then that's a classic case of "everyone wins". A final thought on this model is that increased spend comes with increased risk as well. Agencies who manage millions carry far more risk than agencies who manage hundreds, and that should undoubtedly be reflected in their pricing models.

Alternative Models

An increasingly common option is a combination of a few models, such as a "Flat Rate Based on Spend" model. In this model, there is a combination of percentage of spend and flat rate. There is much value in a model that allows growth in fees without having to consistently hit up clients for more cash. It is rare, almost unheard of, for ZATO to officially raise rates on our clients. This is because naturally over time our work in the accounts will help grow the spend, and our rates naturally raise. It's historically been a healthy, symbiotic relationship for us and our clients since a rising tide lifts all ships.

In this model combination, actual ad spend is used as a rate of growth to evaluate where the flat client fees will be each month. This is a helpful model in positioning it to clients who are allergic to % of spend, even though, ironically, it is still based on their ad spend. The largest negative I have found to this model over the years, is that the jumps between tiers can be significant enough to cause the client to balk to the agreement they "agreed" to a year ago when their fees (and the allotted work) was much lower. In that case, the client on one hand is fair to wish they did not have to pay higher fees, and the vendor is fair to be frustrated that the client is balking at what they previously agreed to.

I have personally found that % of spend increments over a base "flat" rate is an acceptable model as it allows for smaller, incremental growth without significant jumps from one flat rate tier into another.

Before wrapping up this appendix, it is worth noting two random things to get you thinking.

First, at ZATO, we charge a setup fee for the amount of extra work we do in every account we take over. It is worth it to me, since we are actually managing more, digging in with more research, having more communication, etc., in a new account.

Second, consider automated credit card payments, especially if they are prepaid. We utilize a service to charge automated payments to clients, pre-paying the month ahead (then we charge their % of spend overage, if it applies, after the month is completed). For a consultant or small agency, being able to spend less time invoicing and chasing down overdue invoices is worth its weight in gold.

Whatever you do, don't settle for a pricing model that limits your ability to be paid what you are worth. Find a model that works for you, and then be willing to keep thinking about making it better. Maybe if we all keep chasing the perfect model, we'll actually find it.

ABOUT THE AUTHOR

Kirk is the owner of ZATO, his micro-agency focused solely on Paid Search Advertising, and has been knee-deep in advertising since 2009 when he needed a job (any job!) to get him through a Masters in Theology in Louisville, KY. He was named one of the Top 25 Most Influential PPCers in the world by PPC Hero for 6 years in a row, and is known for his book and TEDx Talk titled the same: Stop The Scale.

He is also known for his PPC articles across various industry publications including Shopify, Microsoft, Moz, and MarketingLand. Kirk is an international conference speaker presenting on all things Paid Search (especially Shopping Ads), and has spoken in London, Dublin, Milan, Sydney, as well as many US cities. Kirk has written three books on PPC marketing and agency ownership and loves to teach PPC whenever and wherever he has the opportunity. Kirk currently resides in Billings, MT with his wife, 6 children (yup 6), Trek bikes (he is an avid road biker when the Montana weather cooperates), Taylor guitar, books (mostly audio while he's driving or cleaning the house), and probably not as much sleep as he should be getting.

Printed in Great Britain
by Amazon